A-Z
BIOLOGICAL PHYSICS

A-Z BIOLOGICAL PHYSICS

Prof. Ram Naresh Mahaling

CENTRUM PRESS
NEW DELHI-110002 (INDIA)

CENTRUM PRESS
H.O.: 4360/4, Ansari Road, Daryaganj,
New Delhi-110 002 (India)
Ph.: 23278000, 23261597

B.O.: No. 1015, Ist Main Road, BSK IIIrd Stage
IIIrd Phase, IIIrd Block,
Bangalore - 560 085 (India)
Tel.: 080-41723429
Visit us at: www.centrumpress.com

A-Z Biological Physics

First Edition, 2009

ISBN 978-93-80106-06-9

PRINTED IN INDIA

Printed at Salasar Imaging Systems, Delhi-110035 (India)

Contents

Preface

The fields of Biophysics, Biochemistry and Molecular Biology have developed from Physiological research. Biological Physics make up for a sizeable component of Organic life, in addition to making visible immense diversity in Biological life, Physiology becomes more interesting and viable. A-Z is intended for the biologists for whom a thorough, knowledge of the Physical Forces operating in the Biological world, and a proper perspective on the levels of Organisation and the operations taking place within the Organisms, are absolutely essential.

A-Z BioPhysics provides a working knowledge on the relationships between Physical Properties, Chemical Constitution and Biological Function. The nature of Organisation of matter and the basic Bioenergetics are treated comprehensively, followed by a discussion on the three-dimensional orientations of Macromolecules and Biomembranes, which hold the key to their alignment, location and relegation of duties.

Biophysics often does not have university-level departments of its own, but has presence as groups across departments within the fields of molecular biology, biochemistry, chemistry, computer science, mathematics, medicine, pharmacology, physiology, physics, and neuroscience. What follows is a list of examples of how each department applies its efforts toward the study of biophysics. This list is hardly all inclusive.

Author

Chapter 1

Biological Physics

In biological physics, quantitative aspects and ideas and techniques of physics research are applied to biological problems. Physics has been evolving in the direction of studying more complex systems, and in doing so, has begun to expand its subject area to include living systems.

For example, Professor Joachim Mueller leads an experimental group who are studying the kinetic pathways that occur when proteins misassemble, a topic that has relevance in research on Alzheimer's disease. His lab features a novel approach to fluorescence fluctuation spectroscopy (a process where a fluorescent compound is injected into a cell and studied under a microscope) with the innovation of a high pressure cell that allows spectroscopy over a broad range of pressures. Biological Physics is an interdisciplinary field which applies techniques from the physical sciences to understanding biological structure and function at the molecular level.

In simpler terms we use a range of scientific techniques to try to understand what macromolecules like DNA, proteins, fats and sugars look like, and how they interact to form the biological systems around us. Plants, animals and even seemingly simple organisms like bacteria and archae consist of complex biological networks of chemical signals and molecular interactions, which enable them to do such things as move, respire or reproduce.

Biological Physics is a varied field which draws on biology, physics, chemistry, mathematics, engineering, genetics, physiology and medicine, all with the aim to understand these systems using experiments or theoretical and computational modeling. It is also a young science and is still rapidly developing - bionanotechnology and biosensors are just some of the latest fields to emerge in the last few years. It is also known as biological physics that applies the theories and methods of the physical sciences to questions of biology. Biological Physics apply to theories of the human species and methods of the way we question problems.

Biological Physics research today is comprised of several specific biological studies which neither share a unique identifying factor nor subject themselves to clear and concise definitions. The studies included under the umbrella of Biological Physics range from sequence analysis to neural networks.

Biological Physics is also concerned with creating mechanical limbs and nanomachines to regulate biological functions, although currently these are more commonly referred to as belonging to the fields of bioengineering and nanotechnology respectively. Biological Physics typically addresses biological questions that are similar to those in biochemistry, but the questions are asked at a molecular level.

Traditional studies in biochemistry and molecular biology are conducted using statistical ensemble experiments, typically using pico- to micro-molar concentrations of macromolecules. Because the molecules that comprise living cells are so small, techniques such as PCR amplification, gel blotting, fluorescence labeling and in vivo staining are used so that experimental results are observable with an unaided eye or, at most, optical magnification. Using these techniques, researchers in these subjects attempt to elucidate the complex systems of interactions that give rise to the processes that make life possible.

By drawing knowledge and experimental techniques from a wide variety of disciplines, biophysicists are able to indirectly observe or model the structures and interactions of *individual* molecules or complexes of molecules. Biological Physics is the application of physics to biological systems.

This interdisciplinary field is quickly growing in the modern scientific community. The Department of Biological

Physics, is a centre of drug discovery and clinical proteomics. It has combined the fields of structural biology, bioinformatics and proteomics seamlessly. The goal of modern research in Drug Discovery is to develop drugs that will act in a specific way with minimal side effects and also are demonstrably better than the existing therapies. The conventional approaches of Drug discovery render it a long and an expensive process.

They require screening of hundreds of thousands of samples before reaching some potential compounds with desired properties.

By some estimates, it takes dozens of years and millions of dollars. However, with the advances in protein structure determination, structure based drug design has emerged as a powerful tool for developing new drugs with specific properties and minimal side effects. This technique is usually faster than the conventional methods. In structure based drug design, the three-dimensional structure of a drug target interacting with small molecules is used for drug discovery.

Structure-based drug design represents the idea that one can see exactly how the ligand molecule interacts with its target protein. The designed compounds that have affinities in the acceptable pharmacological range can be further processed for other biological assays and clinical trials.

CHALLENGES FOR PHYSICS

Since physics forms the basis of the life science evolution, the visionary guidance and assistance of physicists who have enjoyed a versatile training will be needed. Perhaps the most easily recognizable example of this need for an interdisciplinary approach is the astounding revelation of the human genetic code, the self-organized plan according to which self-reproducing entities, such as cells, form complex organisms.

The underlying interactions are governed by physical principles. An equally important challenge to physics is to clarify the way in which ensembles of cells, cells as individual entities, and their molecular constituents, function in their respective surroundings.

Biological Physics has thus become a central theme of the physics department, offering young students unique opportunities for study and ample openings for top-level research. Excellent job opportunities and the dynamics of Munich as a centre of biotechnology with its many startup companies underline the attractiveness of the Biological Physics programme of the TU Munich (TUM). The role of physical methods in life sciences is manifested by modern techniques such as ultrasound, positron emission, X-ray and nuclear magnetic resonance tomography, light and electron microscopy, laser spectroscopy, X-ray structure analysis, and electrophysiological techniques such as patch clamp.

These techniques have benefited from more than half a century of research in physics and are the result of combining classical instrumentation with computational physics. The discoveries of these new physical methods have triggered off dramatic progress in life sciences. At the same time there are also numerous examples originating from biology that have inspired new developments in physics. The most prominent one is the discovery of the general energy conservation law by Robert Meyer and Hermann von Helmholtz and the theory of Brownian motion by Albert Einstein.

Einstein's ingenious interpretation of the observation of the botanist Robert Brown that seeds perform random walks in water influenced the development of modern physics at the beginning of the last century nearly as much as Planck's equation describing black body radiation. Nowadays, physicists do not content themselves with being the designers of new instrumentation but strive for a more active role in the search for universal physical principles governing the assembly and function of biomaterials.

In order to be successful, it is absolutely necessary that physicists accept the complexity of biomaterials and become familiar with the principal questions of biology. The physicist's capacity to unravel universal laws governing complex processes or to develop new measuring techniques to test laws predicted by theory is urgently needed in life sciences. As early as 1980, to meet the challenges of modern life sciences, the physics department of the TUM decided to strengthen its resources in the field of Biological Physics.

There are now 5 full professors at the TUM, complemented by 3 adjunct professors, all working in the field of Biological Physics. Recently, a junior professorship and a junior group funded by the Emmy Noether programme were established so as to broaden the scope of the research activities.

The TUM was, in fact, the first university in Germany to introduce a full Biological Physics programme. The combination of experimental and theoretical studies on single molecules, living cells, and cellular networks is a unique strength of the biophysical research efforts of the physics department. The goal is not only to obtain a detailed understanding of biophysical principles but also to develop new tools with which to study and quantify biological processes. It is this combination of different approaches and foci that makes the study of Biological Physics at the TUM an outstanding experience.

The Biological Physics research groups of the physics department engage in a very broad spectrum of activities and are complemented by several special areas of research.

RESEARCH ACTIVITIES OF BIOLOGICAL PHYSICS

Physics Department E13

Other chairs of experimental physics are also engaged in biophysical activities.

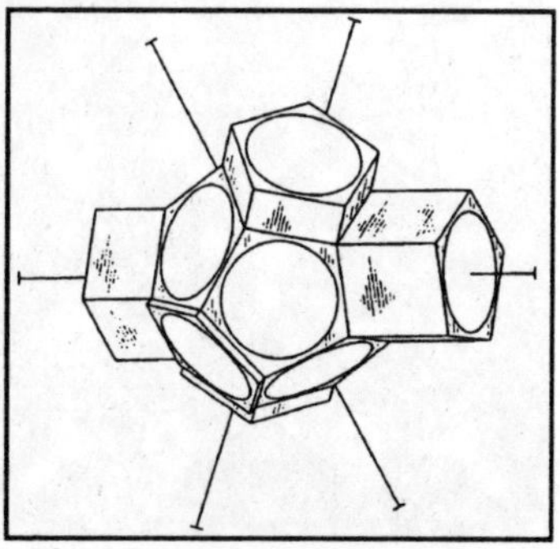

Fig. Neutron Spectrometers

One example is E13 where protein dynamics is investigated by inelastic neutron and light scattering techniques providing information about viscolelastic properties from picoseconds to seconds. New neutron spectrometers are developed to address biological questions.

Physics Department E14

Proteins are never at rest, not even at the lowest accessible temperatures. They constantly change their structures. Although these changes are small, the respective dynamics can be precisely measured by means of the hole burning technique. Hole burning is a non-linear optical technique for high resolution spectroscopy and is widely used by the E14-group to investigate protein dynamics and the associated laws of structural motions. Holes are also used to measure the electrostatic properties of proteins as well as their elastic properties. They are likewise used to unravel the various contributions to the interaction between proteins and prosthetic groups.

Physics Department E17

Molecular Biological Physics: proteins are the bricks of life and are involved in nearly all functions in living beings. They are at the centre of interest of E17. Proteins have a complex architecture.

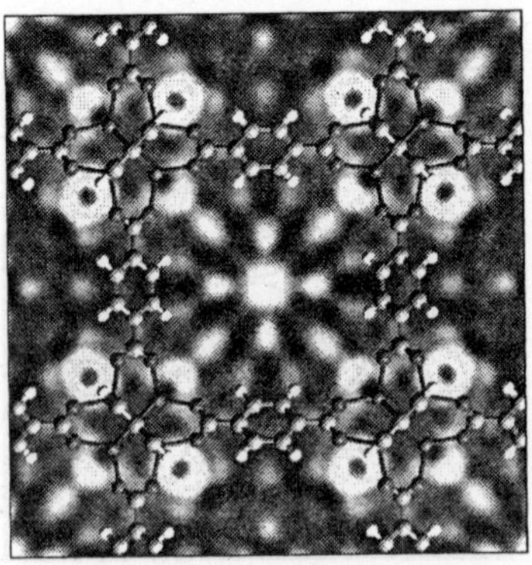

Fig. Neutron Scattering

However, their function is also closely related to their dynamics. X-ray structure analysis yields information on their structure as well as on structural fluctuations. The functionally important hydrogen positions are determined by neutron scattering.

The use of synchrotron radiation at DESY in Hamburg enables us to determine the structural environment of metal ions in proteins. Mössbauer absorption spectroscopy provides us with direct access to dynamics on a time scale faster than 100ns.

Nowadays it is possible to measure the Mössbauer spectra of 57Fe containing proteins in the time domain using nuclear forward scattering of synchrotron radiation. Phonon densities in proteins are determined by the "Phonon Assisted Mössbauer Effect".

Applications in medicine: in collaboration with the "Hals-Nasen-Ohrenklinik" (ear, nose and throat clinik) and the "Poliklinik" (outpatients' department)" E17 is attempting to directly deposit chemotherapeutic agents in tumors by means of "magnetic drug targeting". Spectroscopic investigations of the Cu(II) binding site of the cellular prion protein are also being performed to obtain information on BSE.

The use of synchrotron radiation at DESY in Hamburg enables us to determine the structural environment of metal ions in proteins. Mössbauer absorption spectroscopy provides direct access to dynamics on a time scale faster than 100ns.

Physics Department E22

The group of physicists researching into cellular biological systems applies general physical principles, e.g., hydrodynamics, statistical thermodynamics, and elasticity, to complex biomaterials such as cell membranes and intracellular networks. The research addresses key questions such as what are the mechanical properties of single cells, how do living cells regulate their mechanical properties locally and what are the physical principles of bio-adhesion?

The applied Biological Physics groups are working on the bio-functionalization of semiconductor devices for the development of biosensors in close cooperation with semiconductor physics.

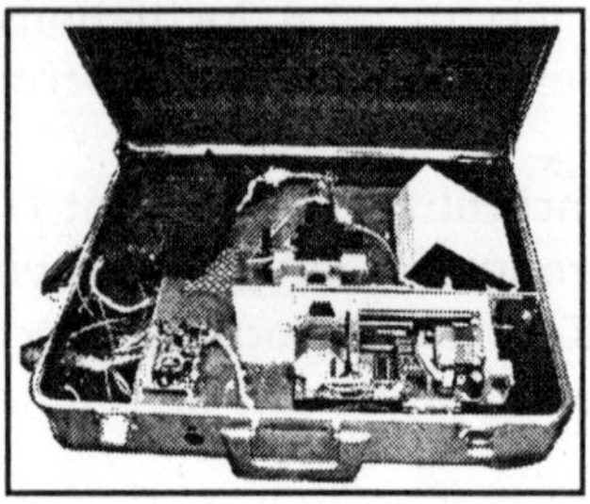

Fig. Biosensors

The questions central to these research efforts are how we can construct a sensitive biosensor, how we can detect a small number of biomolecules on solid surfaces and how can semiconductor devices be combined with biological systems such as cells?

THE THEORETICAL BIOLOGICAL PHYSICS

Physics Department T38

Molecular dynamics studies using computer systems to understand the mechanisms of molecular machines.

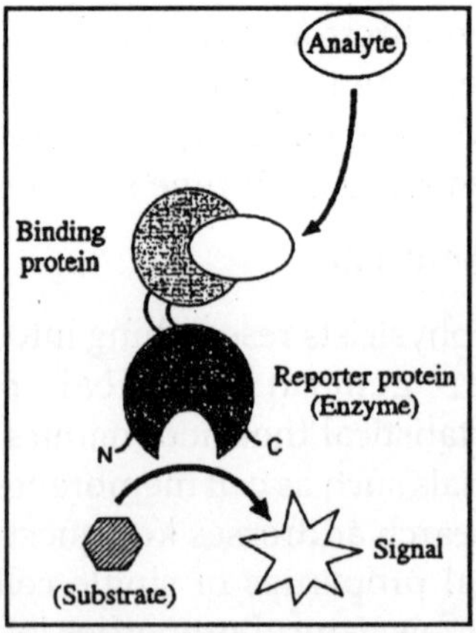

Fig. The Function of Bacteriorhodopsin can be Studied Using Molecular Dynamic Techniques.

The key questions being addressed are how we can understand the dynamics of the elementary processes in biological systems such as in the functional units of photosynthesis?

How can we apply quantum-chemical calculations to electron transfer, proton transfer and conformational changes of the protein?

Physik Department T35

Information processing within biological systems is a fascinating topic for theoretical Biological Physics, especially in neuronal systems where nature has devised 'smart' procedures for detecting and processing sensory input. Highly evolved information-processing being a true biophysical challenge, we analyse performance of the visual, auditory, and other sensory systems of animals with specialized circuitry for computing prey location.

Neurons communicate through conduction of action potentials, or voltage 'spikes', of about 0.1 V amplitude and1 ms duration. Spikes travel along axons as delay lines to synaptic terminals on other neurons. Information is generally stored in synapses. Our work focuses on how neurons represent the outside world, in particular, their dynamics and the way in which simple forms of synaptic plasticity can act to guide and perfect performance, with time coding as the unifying concept.

Dynamics Given the existence of many areas in each of the sensory domains, e.g., about 30 in visual cortex, most of which handle specific aspects of sensory input, a natural question is how they together code a single object. This is called the binding problem. A closely related problem is the dynamics of the brain's feedback, which is at least as strong as the feedforward interaction. Understanding the role of feedback is also essential to understanding attention, an efficient means of reducing what is 'perceived' of the input.

Synaptic Learning Neuronal activity patterns that are to be stored are time-dependent and, hence, spatio-temporal. In the late eighties we proposed the first mathematical formulation of a learning rule for spatio-temporal patterns. A final version of our time-dependent learning rule, meanwhile extensively confirmed by experiment, specifies synaptic learning as a process depending on the arrival time of an input spike with respect to firing of the postsynaptic neuron; a synapse as the place where information is stored.

Time Coding Neural information is encoded through specific temporal patterns of spike activity. Many of nature's most precisely timed events occur in relation to detection of prey or social signaling, behaviours with intense selective pressures for precision of spatial localization. To uncover universal principles, we study representative cases of different localization strategies so as to discern the underlying principles of time coding.

We showed how neuronal coincidence detection in the barn owl's auditory system generates high precision both in neuronal performance and in generating the azimuthal map as a neuronal representation of the outside world. In the sand scorpion, we have discovered a temporal coding mechanism relying on subtraction (instead of addition as in the barn owl) of nervous activity stemming from different mechanoreceptors that detect surface waves in desert sand.

Chapter 2

Importance of Fluid Transport

CARRIER MEDIATED TRANSPORT

Properties of Carrier Mediated Transport

This is a category of processes that accounts for passage through the membrane of large, lipid-insoluble molecules and ions. All use carrier systems, i.e., special carrier molecules are required to be present in the membrane to help in the movement of the transported molecules. These carrier systems all show three properties: specificity, saturation and competition. By specificity, we mean that only certain molecules are transported by the system.

The transport system that moves glucose works only for the D-isomer not for the L-isomer, and it doesn't work for other sugars. Because there are a limited number of special carrier molecules in the membrane, only that number of transported molecules can be accommodated at one time.

Having more present does not increase the rate of transport. For simple diffusion, as concentration increases so does the rate of diffusion and the relationship are linear. However, with mediated transport, the rate of diffusion increases as concentration increases up to a point, and then there is no further increase in rate with increasing concentration.

At this point, we say the system is saturated. Presumably, all of the carrier molecules are being used. A physical example may help to visualize this property. Suppose 75,000 people want to get into Memorial Stadium to watch a football game.

If there is an unlimited number of entry doors, then all of the people can enter at the same time—the rate is maximum.

If you have been there, you know that there is a limited number of doors, and the rate is quite a bit less than maximum. When two or more substances use the same carrier molecule, then the presence of one will slow the rate of transport of the another. They compete for the use of the carrier molecules. This is another obvious consequence of a limited number of carrier molecules, and it is called competition. With competition the rate is slower than.

FACILITATED DIFFUSION

This process behaves much like simple diffusion in that net transport occurs down the concentration gradient for the transported substance. Because it occurs downhill, no external energy is required; the energy required by facilitated diffusion comes from the same source as that for simple diffusion—thermal motion. As in simple diffusion, there is an equilibrium condition for facilitated diffusion.

There will be no net transport when the concentrations on the two sides of the membrane are equal. Glucose can get through the membrane by diffusing through the pores (its diameter is approximately that of the largest ones), but the rate at which it enters most cells is higher than one would expect for diffusion.

For most cells, most of the glucose enters by facilitated diffusion. Once glucose enters the cell it is immediately metabolized to glucose-6-phosphate. So, the intracellular concentration of glucose is always near zero. Thus, there is a continuous gradient toward the inside of the cell, and the facilitated diffusion system will always bring glucose *into* the cell.

However, if we artificially raise the glucose concentration within the cell above that outside the cell and prevent it from being metabolized, then the facilitated diffusion system will take glucose *out* of the cell. Thus, the direction of transport is determined by the concentration gradient, not by some property of the transport system itself. Presumably, the carrier molecule is more likely to attach to the transported molecule where that molecule is in highest concentration. Once the two are attached, a conformational change in the carrier molecule exposes the transported molecule to the inside of the cell, where it is released because of its lower concentration. At least, that is one model of how the process works. Glucose enters most cells by facilitated diffusion. In addition, amino acids and sometimes sodium ions are transported by this mechanism.

ACTIVE TRANSPORT

Sometimes molecules must enter or leave a cell even though they are in higher concentration at their destination. Sodium ions are continuously "pumped" out of all cells, though their concentration is ten times higher outside. We have seen that under these conditions sodium ions have a natural tendency to enter the cells by diffusion down their concentration gradient. Therefore, to move them out of the cell requires expenditure of energy. A form of equilibrium also applies to active transport.

When the concentration increases to the point that particles "leak" into the cell at the same rate that they are pumped out by the active transport system, a steady state or equilibrium condition will exist. Such active transport systems or "pumps," as they are often called, probably only transport ions—charged particles.

Uncharged particles may move by coupling with the ions being transported. Some pumps are capable of moving more than one ion at a time.

Pumps can be electrogenic or non-electrogenic, depending upon the number of charges transported in each direction across the cell membrane.

If the sodium pump were to pump more sodium ions out of the cell than it pumps potassium ions in, there would be a net outward current. This current, through resting membrane, would slightly hypopolarize the cell membrane (electrogenic). Conversely, if the same number of sodium and potassium ions were pumped, then there would be no net current and no change in membrane potential (non-electrogenic). One model of the active transport process has the carrier molecule with a higher affinity for sodium ions on the inside of the cell.

Attachment of a sodium ion causes a conformational change in the carrier molecule, exposing the sodium ion to the outside of the cell, where the affinity of the carrier molecule for sodium is smaller, and the sodium is released.

Co-Transport

When a substance moves from a region of higher concentration to a region of lower concentration, as it does during diffusion, it gives back the energy that was supplied to create the concentration gradient in the first place. That energy can be used to move other substances, even up their concentration gradients. This is the essence of co-transport or ion-coupled transport

SYMPORTS

In a symport, the two substances being transported move in the same direction across the membrane. In the intestine, sodium ions diffuse into the lining cells by diffusion along the normal outside-to-inside gradient. There is a carrier system in the cell membrane which couples this movement to an inward movement of glucose molecules. Glucose could not be transported actively because it has no charge, but it can move by this process, even up its concentration gradient.

It is for this reason, that there is normally no glucose in the feces—it all gets removed by this symport. This process does not require external energy. The sodium pump can be poisoned, and it will still continue as long as a sodium concentration gradient remains.

However, some gradient must be present; so the entering sodium must be removed from the cells. Apparently, sodium pumps in these luminal cells are sequestered to the blood side of the cells; all entering sodium is pumped out of the cells into the blood, not back into the intestinal lumen. Glucose is removed from the cell by facilitated diffusion, and water follows along with both sodium and glucose in order to keep its concentrations in equilibrium.

ANTIPORTS

A similar mechanism can exchange substances across membranes. Several substances are known to be exchanged for sodium ions. In all cells, there exists a H^+-Na^+ antiport. In cardiac muscle, there is a Ca^{2+}-Na^+ antiport. The latter is blocked by both digoxin and ouabain, which are known to inhibit ATPase activity.

When the antiport is blocked, calcium concentration rises within the cells, strengthening the force of contraction. Antiports do not have to involve sodium ions. In erythrocytes and elsewhere, there exists a Cl^--HCO_3^- antiport. Presumably, this plays some role in buffering within the cells that have it.

PHAGOCYTOSIS AND PINOCYTOSIS

Sometimes water-soluble substances too large to pass through the pores must get into or out of cells. In addition, there are times when the contents of vesicles must be released all at once. These take a special mechanism.

EXOCYTOSIS

During exocytosis, a substance is relased from the cell. When a substance is to be released, during synaptic transmission (the communication between two nerve cells), the substance is usually stored within a secretory vesicle.

The release process involves migration of the vesicle to the cell membrane, fusion of the vesicle membrane with the cell membrane, and rupture of the point of junction. The substance contained within the vesicle then finds itself within the extracellular fluid.

ENDOCYTOSIS

To take a substance into the cell, the process is simply played out in reverse, and it is called endocytosis. The substance moves to the cell membrane or the cell moves to the substance.

A portion of the membrane surrounds the substance, and the substance containing portion "pinches" off to form an intracellular vesicle.

This special mechanism is called phagocytosis or pinocytosis, which translate to "cell eating" and "cell drinking." The former is applied to movement of particulate material, the latter to fluids. In exocytosis the vesicular membrane actually becomes a part of the cell membrane. Conversely, in endocytosis a portion of the cell membrane is lost in formation of a vesicle. In this way, the cell membrane is constantly turning over. Clearly, there must be a balance between endocytosis, exocytosis—the formation of new membrane and the degradation of old membrane—if the cell is to maintain its size.

SIMPLE DIFFUSION

Let's examine the behaviour of ions in solution in some detail. If we place some salt, NaCl, into one of two compartments separated by a removable partition, and add equal volumes of water to both compartments, we can determine how solutes move in solutions.

Within a few minutes after removal of the partition, the solutions in both compartments will contain salt, and the solutions will have the same concentrations. The salt must have moved throughout the compartments; this movement is called diffusion.

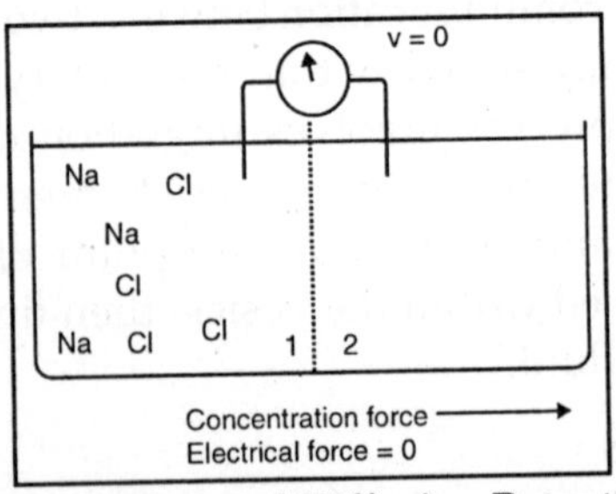

Fig. Diffusion and Diffusion Potentials.

A container, with a removable partition, contains water. Ordinary table salt, NaCl, is added to side 1 and then the partition is removed. After a time, NaCl is found in equal concentrations of both sides. Molecules in solution at temperatures above absolute zero are in random motion, so-called Brownian motion. The direction of motion of any one molecule can be changed by collision with another molecule or the side of the vessel, but once in motion the molecule will travel in a straight line until a collision occurs.

The higher the temperature, the greater is the velocity of movement; the higher the concentration of the solution, the greater is the likelihood of a collision. Some collisions will result in molecules moving toward the interface between the two compartments and crossing it.

The probability that a given molecule will cross the interface is therefore proportional to both temperature and the concentration of the solution. Thus, the flux from one compartment to the other, J, is given by:

$$J = kC$$

where k is a constant for a given temperature and C is the concentration in the compartment from which the flux is occurring.

The side where the salt was put is numbered 1; the other side is numbered 2. Therefore, the flux from side 1 to side 2, J_{12} is:

$$J_{12} = kC_1$$

where C_1 is the initial concentration of salt on side 1. As soon as some molecules have crossed the interface, the concentration on side 2 begins to increase, and there will be an increasing likelihood of a molecule recrossing the interface back to side 1. This probability is related to temperature and concentration—this time, to the concentration on side 2; the back flux, J_{21}, is given by:

$$J_{21} = kC_2$$

The net movement or net flux, J_N, between the two compartments at any time is simply the difference between the two one-way fluxes, J_{12}, and J_{21},

$$J_N = J_{12} - J_{21} = kC_1 - kC_2 = k(C_1 - C_2)$$

Diffusion is simply a net flux. The net flux is positive when $C_1 > C_2$, so the diffusion is from a region of higher concentration to a region of lower concentration, down the concentration gradient. The net flux is proportional to C_1-C_2, or proportional to the concentration difference between the two compartments.

Equilibrium is defined as the condition of no net flux, $J_N = 0$. When $J_N = 0$, $J_{21} = J_{12}$ in equation. Expressed in words, the equation means simply that the two one-way fluxes are equal at equilibrium, but neither flux need be zero. Even at equilibrium, molecules can move from one compartment to the other, but equal numbers must move in both directions. Only the net flux need be zero. (This condition is sometimes called a "dynamic equilibrium.") Equation 4 can be solved for $J_N = 0$ as follows:

$$0 = J_N = kC_1 - kC_2, \text{ or } kC_1 = kC_2, \text{ or } C_1 = C_2$$

Thus, equilibrium will occur when the concentrations in the two compartments are equal. Until equilibrium is reached, there will be a net flux from side 1 to side 2 that is proportional to the concentration difference.

If we introduce a membrane, permeable to NaCl, between the compartments and repeat the experiment, the result will be the same. The net flux will be proportional to the difference in concentration between the two compartments with the constant of proportionality equal to the permeability coefficient, p, of the membrane for NaCl or $J_N = p(C_1 - C_2)$. The only real difference will be that the process occurs more slowly.

*Diffusion Speed:*The speed of diffusion of a substance may be calculated from the equation:

$$x^2 = 2Dt$$

where x is the diffusion distance, t is the diffusion time and D is the diffusion coefficient. Taking a simple example will illustrate that diffusion is a slow process. Let $D = 10^{-5}$ cm^2/sec. Then we can calculate t for different values of x:

x	t
1 micrometer	0.5 msec
100 micrometers	5 sec
1 cm	14 hr

Clearly, diffusion is rapid if the distance to the target is 1 micrometer or less; it is marginal if the distance is 100 micrometers. Many insects breathe through their skins, the air simply entering the cells directly. For them this is possible because they are small and the cells are located near a repository for air. Clearly, this mechanism would not work in a human. The distance from the skin surface to many organs is more than 1 cm, requiring an inordinately long time for gas exchange.

A similar situation surrounds supplying nutrients to body tissues. Many of them are located more than 1 cm from the source of the nutrients. It is for this reason that we have lungs and a circulatory system.

ENTRY INTO CELLS

Lipid soluble substances can enter a cell by dissolving in the lipid portion of the membrane and diffusing through it. The greater the lipid solubility, the more readily a molecule will pass through the membrane, i.e., the greater will be the flux for a given concentration difference. The oil-water partition coefficient (solubility in oil/solubility in water) gives a useful measure of lipid solubility. Molecular size is not very important for lipid soluble molecules although very large ones enter cells a bit less readily.

Examples of lipid soluble molecules that can pass through the membrane include: oxygen, carbon dioxide, steroid molecules, and anesthetic agents. Water soluble, lipid-insoluble substances by definition cannot pass through membranes by dissolving in the lipids. How do they get through? There must be water-filled channels or pores spanning the membrane through which these substances may diffuse. These will admit substances up to about 200 Daltons in molecular weight, i.e., about the size of a glucose molecule.

These pores are of different sizes, some as small as 4 Angstroms and as large as 8 Angstroms. Most substances cannot pass through pores because of their large size, especially when hydrated. The number of pores in an area of membrane is difficult to estimate, partly because they appear to be continuously recycled.

Only 0.01% of a cell's membrane need be occupied by pores to account for the observed permeability. Substances that can pass through these water-filled channels include: water, urea, some dissolved gases, and ions such as Na^+, K^+ and Ca^{2+}.

Not all ions get through the membrane with equal ease. This is partly a function of the radius of hydration, but the structure of the channel also plays an important role.

ION TRAPPING

Diffusion of a substance depends only upon the concentrations *of that substance*. Thus, the presence of chloride in a solution of table salt, has no effect on the diffusion of sodium. Likewise, the presence of a metabolite of a substance or the ionic form of a substance has no effect on the diffusion of the substance itself. Glucose is metabolized to glucose-6-phosphate immediately upon entering a cell. The presence of this metabolite does not affect further diffusion of glucose.

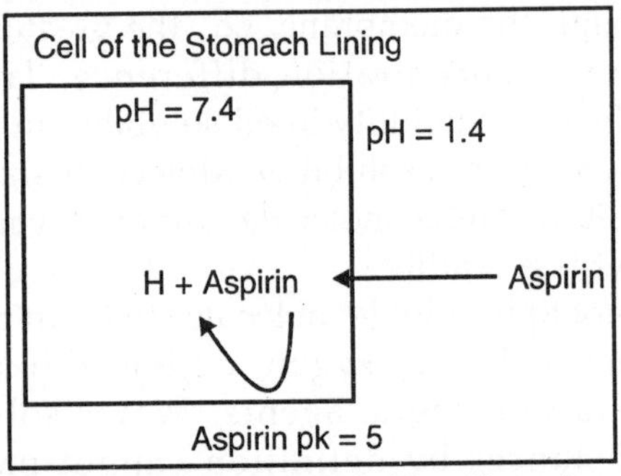

Fig. The Phenomenon of Ion Trapping.

The rectangle is supposed to indicate a cell in the stomach lining. Aspirin has a pK=5. At the higher pH inside the cell, aspirin ionizes and then cannot leave the cell.

This property gives rise to a phenomenon known as ion trapping. Aspirin in the stomach, which has pH=1.4, exists mainly in the non-ionized (uncharged) form. In this form, it can enter the cells of the stomach lining. Once it enters the cells, which have pH=7.4, the aspirin is ionized because it has a pK=5, and, in this form, it cannot leave the cell. Ionized aspirin is not the same as non-ionized aspirin as far as diffusion is concerned.

Therefore, the concentration of non-ionized aspirin inside the cell remains low, and it continues to enter the cell until it saturates and precipitates as crystals. These may actually rupture the mucosal cells and lead to gastric bleeding. One way to prevent this is drink a lot of water with aspirin in order to lower its concentration inside the stomach. Ion trapping can be exploited for beneficial purposes as well. Aspirin poisoning can be treated by raising the pH of the urine to trap the ionized form there. It will be voided with the urine.

OSMOSIS

Water is a substance just like sodium or potassium or glucose. It will move down its concentration gradient by simple diffusion until equilibrium is reached, just like these other substances. In addition, water is subject to "osmosis," a bulk flow of molecules (like water moving through a tube). Diffusion of water comes about because of differences in water concentration; osmosis comes about because of differences in hydrostatic pressure. When both forces are present, the hydrostatic force will dominate.

Diffusion of water has no effect upon the diffusion of other substances, but during osmosis, dissolved substances can get swept up in the stream of water and moved faster than would be expected from their diffusion alone. This is called "solvent drag." Osmosis requires that the cell membrane be semi-permeable, meaning that not all substances can pass through it. The membrane must be permeable to water (all are), but it must be impermeable to some other substance.

Refer to particles that can pass through the membrane as "penetrating" and to those that cannot as "non-penetrating." Pure water has a concentration of 55.5 moles/liter. Anything added to it lowers the water concentration. It is easier to express water concentration in terms of solute concentration, but all "osmotically active particles" must be considered. Thus, water concentration is expressed in osmoles, the number of moles of dissolved *particles* per liter of solution. A 1 M solution of glucose is, therefore, a 1 osmolar solution.

Similarly, a 1 M solution of urea is a 1 osmolar solution. But, a 1 M solution of NaCl is a 2 osmolar solution because NaCl dissociates into two particles in solution. Furthermore, a 1 M solution of Na_2SO_4 is a 3 osmolar solution. Bear in mind that the higher the osmolarity, the lower the water concentration. Therefore, water will move from regions of *low* osmotic concentration (high water concentration) to regions of *high* osmotic concentration (low water concentration). Consider a container, divided into two compartments by a semipermeable membrane through which water penetrates, but some solute does not.

Add some non-penetrating solute to the left compartment. Because the solute dissolves completely, there will be essentially no change in volume of the left compartment. After some time, the level of the water in the right compartment will have fallen, that in the left will have risen. The only possible explanation for this occurrence is that water has moved from right to left.

This makes sense from what we know of diffusion. Initially, the concentration of water on the right was 55.5 M or 100%, that on the left was something less than this value. Water has diffused down its concentration gradient. But, consider a different situation.

Two chambers of a reservoir of water are separated by a semipermeable membrane as before. The left chamber contains some concentration of a non-penetrating solute and a piston that can move. As water enters the left chamber it pushes the piston to the left.

The pressure gauge at the top of the left chamber does not change its reading—the force exerted by the increase in water is used to push the piston. If the piston is fixed, then it cannot move and the pressure measured by the gauge will increase. This pressure is called the osmotic pressure.

The osmotic pressure of a solution may be calculated from Van't Hoff's Law:

$$P = iRTc$$

where i is the number of ions formed by dissociation, R the ideal gas constant, T the absolute temperature and c the molar concentration. It can be seen that a 1 milliosmolar concentration difference will yield a 19 mmHg osmotic pressure. Osmotic pressure is one of the parameters that determines whether water enters or leaves the blood in blood vessels. Some people use the term "oncotic pressure" or "colloidal osmotic pressure" in this application.

The blood contains large protein molecules (e.g., albumin), which are non-penetrating. The greater their concentration, the greater will be the tendency for water to enter the capillaries.

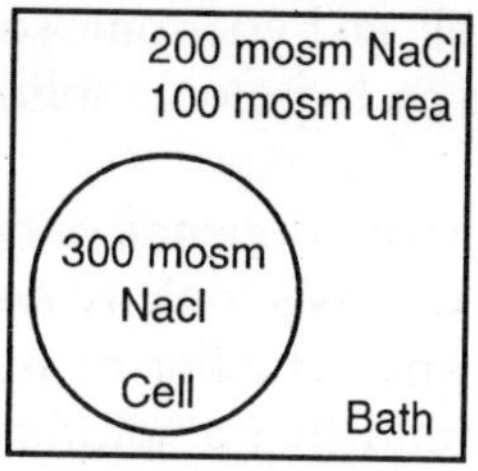

Fig. Cell Found in a Large Bath. Components of the Bath and Cell are Indicated.

Osmotic pressure is also a major determinant of the size or volume of individual cells. A cell placed in solution may change size depending upon what is in the solution. Obviously, a physician does not want to change the size of cells in a patient's body.

Therefore, it is important to exercise care when administering fluids to a patient. A solution whose osmotic concentration is greater than that of cytoplasm is called an hyperosmotic solution. A solution whose concentration is less than that of cytoplasm is called an hyposmotic solution.

The final possibility is an isosmotic solution. Just knowing the osmolarity of a solution is not sufficient to determine what will happen to cells. The osmotic concentration of the cell's cytoplasm is 300 milliosmoles/liter.

The osmotic concentration of the solution into which it was placed is also 300 milliosmoles/liter, 200 from NaCl and 100 from urea. With equal osmotic concentrations and, therefore, equal water concentrations, you might think that the cell would stay the same size—water would neither enter nor leave. However, urea is a penetrating particle; it easily enters and leaves cells.

For living cells, NaCl is non-penetrating because of the sodium pump (chloride follows sodium to preserve electrical neutrality). The urea will enter the cell down its concentration gradient and carry water with it. That will cause the cell to swell. It is obvious that osmolarity does not indicate uniquely what will happen to cell volume in solution.

A different terminology gives us that unique determination. An isotonic solution is one in which a cell will neither shrink nor swell, an hypertonic solution is one in which a cell will shrink, and an hypotonic solution is one in which a cell will swell.

Obviously, osmolarity depends only upon the number of particles in solution, whereas tonicity depends both upon the number of particles and whether or not they penetrate the membrane. Bear in mind that a solution of only penetrating particles will effect no change in volume of a cell.

Chapter 3

Human Eye

VISION

Photoreception is a particularly important sense for most primates, including man, but it is not unique to primates or even mammals. Even mollusks have photoreceptors, but one may question whether they possess vision in the same sense as we have it.

Most objects reflect light, and because light travels at high speed, it is possible to nearly instantly assess their shape, size, position, speed, and direction of movement.

The light rays emanating from an object are gathered and focused onto an array of photoreceptors. Activities generated in the different photoreceptors by the light interact to produce a two-dimensional representation of the object which is transmitted to the brain. The brain then reconstructs a three-dimensional representation using information received from the two eyes. The end-products of the activity of the visual system are sensations that represent the object and its surroundings.

These sensations can be used to guide our immediate behaviour, or they can be stored for future reference. Visual sensations contain a great deal of information, and understanding these complex phenomena is no simple matter. The best place to begin the study of vision is at the eye itself.

It consists of two fluid-filled chambers separated by a transparent structure, the lens. Nearly the entire eye is covered with a tough, fibrous coating called the sclera that is modified anteriorly to form the transparent cornea.

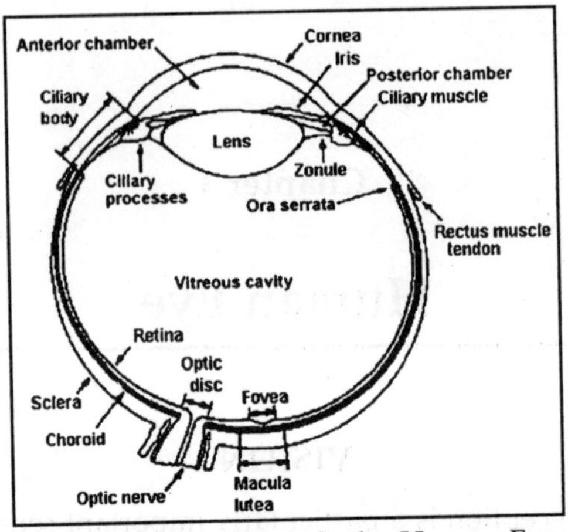

Fig. A Section Through the Human Eye Illustrating the Major Structures.

The human cornea is about 12 mm in diameter, about 0.5 mm thick in the centre and 0.75 to 1 mm thick on the edge, and it is made of the same collagenous connective tissue substance as is the sclera, but the fibres of the cornea are oriented in parallel arrays that let light pass through with minimal scatter, whereas fibres of the sclera are random and light rays are scattered when passing through. The result is that light passes easily through the cornea, but not through the sclera.

Lining the inside of the posterior two-thirds of the sclera are two membranes: the choroid, a pigment layer containing the vascular supply for the eyeball as well as mechanisms for maintaining the integrity of the photoreceptors, and the retina that contains the photoreceptors and other neural elements essential to our visual process. The fine structure of the retina will be considered in detail later. The human lens is about 11 mm in diameter and 3.5 mm thick at its thickest point, and it is suspended in place by zonule fibres that attach to the ciliary process anterior to the retina.

A set of smooth muscle fibres, the ciliary muscle, lie between the ciliary process and sclera. Just anterior to the lens is a pigmented structure called the iris, that is like the diaphragm on some cameras in that it has a hole in the centre of variable aperture, the pupil. The pupil is surrounded by two sets of muscles, one that encircles the aperture, the sphincter pupillae, and one that runs radially out from it, the dilator pupillae. The anterior chamber of the eye is filled with aqueous humour, a watery fluid of low protein content that is formed from plasma.

The vitreous cavity contains a gelatinous substance, the vitreous or vitreous humour. In many people the vitreous is not completely clear, but contains particulate matter that is not transparent. This material may be stationery or may float around, producing "spots before the eyes," the floating variety being called "floaters." At this point, we should spend just a moment reviewing the properties of light, which is a form of electromagnetic radiation. Light is radiant energy emitted by matter when it is at high temperature, when it is electrically excited, and when it is undergoing certain chemical or physical processes. Early on, physicists attempted to account for the properties of light as a stream of particles emitted by a luminous object.

Later, they attempted to account for them in terms of waves, but light has properties both of waves and of particles. In any case, light behaves as if it has characteristic wavelengths that are measurable. The total spectrum of wavelengths of electromagnetic radiation is quite broad, ranging from 1 nanometer (10^{-9} m, abbreviated nm) to over 10^4 m.

Some other animals have a broader sensitivity than man, being able to see in the ultraviolet region (less than 400 nm), but not even these animals see the whole light spectrum, 380-760 nm.

Light travels in straight lines from its source unless it passes through a high intensity electro-magnetic field or from a medium of one refractive index into a medium of another. In the first case, the light is bent along a curved path. When light passes from one medium into another of a different refractive index, the direction of travel of the light beam is changed at the interface. Psychologically, light has three aspects we need to consider. It has hue, a quality correlated with wavelength and analogous with pitch in hearing.

It has saturation, a quality related to homogeneity or purity of wavelength and comparable to timbre or freedom from noise in audition. The colour of a light is a function both of its hue and saturation. Finally it has brightness, a quantitative measure of the intensity of the light and somewhat analogous to loudness in hearing. The intensity of light is dependent upon the amplitude of its waves or alternatively the number of photons (particles of light).

OPTICAL FUNCTIONS OF THE EYE

The process of light reception and transmission has a number of parts. Consider the optical functions of the eye: accommodation, the process of forming a focused image, and control of the amount of luminous flux at the retina.

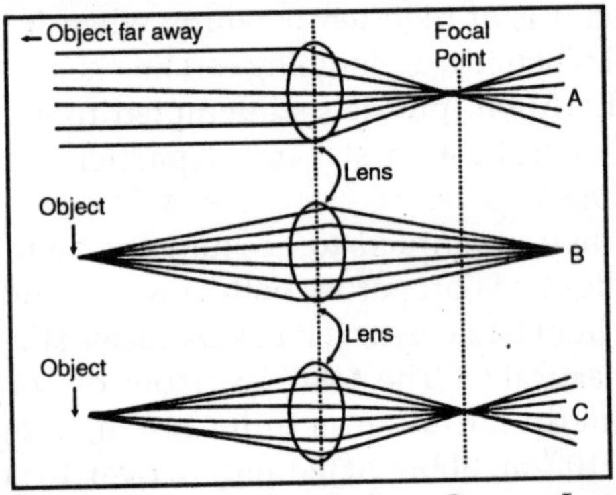

Fig. Convergence of Light by a Convex Lens.

A. Light: From an object far away is brought to focus at the focal point of the lens.

B. Light: From a nearer object is brought to focus at a point farther way than the focal point of the lens.

C. The radius of curvature: The lens must be decreased to bring the light to focus at the focal point of the lens in A.

Accommodation

A convex lens like that in the eye has the property of bending penetrating rays of light toward its axis. This property results from the fact that light travels more slowly in some materials, e.g., water and glass, than in air (and more slowly in air than in a vacuum). The amount of slowing in a material is specified by its refractive index, which is the ratio of speed of light in a vacuum (the speed of light in air is slightly less, but very close) to the speed of light in the material (refractive index = speed in vacuum/speed in material).

If light passes from one medium to another at an angle to the interface between them, the rays are bent, where parallel rays of light emanating from an object far away are bent as they pass through the convex lens in such a way that they converge onto a point in space. Where they converge depends upon the strength of the lens which is specified by its focal length. The focal length of a lens is the distance from the centre of the lens to the point of focus of light from a distant object. Focal length is a function of the refractive index of the lens material and the radius of curvature of the lens.

The larger the radius of curvature, i.e., the less spherical the lens, the longer the focal length is. A convenient way to specify the focal length or strength of a lens is in terms of the diopter, the reciprocal of the focal length of the lens expressed in meters. As you know, there are two kinds of lenses, convex lenses (converging lenses), and concave lenses (diverging lenses). With a convex lens the image of an object is formed on the side of the lens opposite from the object.

With a concave lens no actual image is formed, but physicists identify a virtual image that lies on the same side of the lens as the object. Because the images lie on opposite sides of the lenses in the two cases, convex lenses are assigned positive diopter values (presumably for their real images) and concave lenses are assigned negative diopter values. A convex lens with a focal length of 0.5 meter is a +2 diopter lens, whereas a concave lens with a focal length of 0.33 meter is a -3 diopter lens.

The higher the absolute value of the diopter rating of a lens, the stronger it is, i.e., the more it converges or diverges light that enters it. When objects are close, the distance from the lens to the point of focus varies with the distance to the object. Thus, the light from a source farther away is brought to focus closer to the lens than that from a source nearer the lens. In the eye, the distance from the lens to the retina is fixed so that in order to focus the light from a near source and also light from a far source the strength of the lens must be increased (i.e., the lens made more spherical), a process called accommodation.

The lens is a flexible structure that, when released from the tension of the zonule, becomes nearly spherical in shape. Normally the lens is under considerable tension and thus is flattened, a condition appropriate for focusing objects at a distance. Under these conditions, light from near objects is brought to focus behind the retina but is unfocused on the retina, and the images of them are blurred. Because of the way in which the ciliary muscle is attached, its contraction results in a relaxation of the tension on the lens, causing the lens to become more curved (i.e., more spherical), thereby bringing nearer objects into focus on the retina.

By grading the amount of contraction of the ciliary muscle and therefore the tension on the lens, the nervous system can focus objects located from about 9 cm from the eye to infinity. Images closer than 9 cm cannot be focused by the human eye.

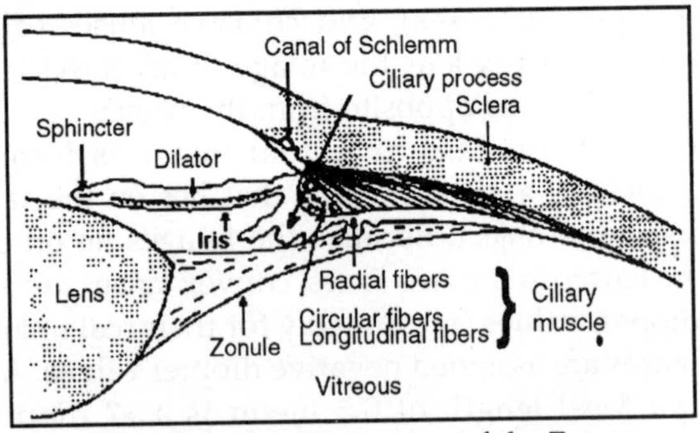

Fig. The Anterior Portion of the Eye.

The arrangement of the lens, the zonule, and the ciliary muscles is illustrated along with the position of the iris and the arrangement of the muscles asociated with it.

Control of the ciliary muscle is exercised by the autonomic nervous system which uses information obtained from the retina itself to adjust the focus of an image. Presumably, the pathway for producing accommodation must involve the retina, optic nerve, lateral geniculate nucleus, visual cortex, superior colliculus and finally the oculomotor nucleus, relaying in the ciliary ganglion to reach the ciliary muscle.

Participation of visual cortex is inferred from the fact that accommodation is normally accompanied by pupillary constriction (miosis of accommodation) and convergent eye movements, in conjunction with the observation that stimulation of visual cortex results in pupillary constriction. The nerve fibres supplying the ciliary muscles are parasympathetic postganglionic fibres and cholinergic (they use acetylcholine as a transmitter substance).

Transmission through the cholinergic junction is readily blocked by atropine, which produces relaxation of the ciliary muscles so the lens is flattened and focused for distant objects. Ophthalmologists frequently relax the ciliary muscles to test the eye for refractive errors, but they use a shorter acting parasympathetic blocking agent, such as tropicamide. They do this in order to put the eye in a standard condition for comparison with a standard normal eye. Actually, the lens is not strong enough by itself to focus a parallel beam of light on the retina.

The cornea also acts as a lens (of fixed strength) and accounts for about 75% of the refractive strength of the optical system. The total refractive power of the eye is about +60 diopters. The cornea accounts for about +45 diopters, the lens only +17 diopters, and the remainder of the +60 is accounted for by the liquid-tissue interfaces. The lens has a higher refractive index than the cornea (1.386 versus 1.376), but it is surrounded by fluids with refractive indices much higher than that of the air (1.000) at the anterior boundary of the cornea. It is the ratio of the refractive indices at an interface that determines the refractive power of that interface.

The greatest ratio, and therefore the greatest refraction, occurs at the air-cornea interface. With accommodation, the eye gains another +10 to +12 diopters depending upon the person's age. Removal of the cornea requires a strong convex lens (about +45 diopters) to correct for the loss in refractive power, but a weak man-made lens will correct for loss of natural lens function. The real problem in loss of lens function is the inability to refocus from near vision to far vision (i.e., to accommodate), a fine adjustment that the lens provides.

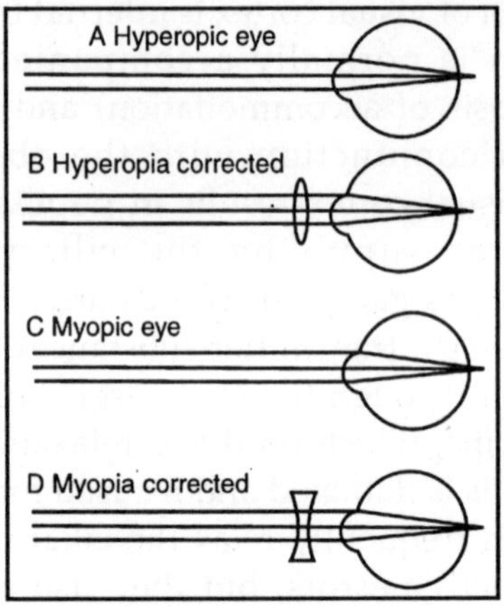

Fig. Errors of Refraction.

A. The Hyperopic Eye: Convergence of parallel light behind the retina.

B. The Hyperopic Eye: Corrected with a convex lens.

C. The Mopic Eye: Convergence of parallel light in front of the retina.

D. The Myopic Eye: Corrected with a concave lens.

ERRORS OF REFRACTION

With increasing age, the lens loses some of its elasticity so that even though the ciliary muscles contract and release tension on the zonule, the lens cannot round-up as it should. The result is that near objects cannot be properly focused, calling to mind the image of a man in his 40's or 50's (the age at which this condition becomes apparent) holding a book at arms length in order to read the small print.

The nearest point at which the eye can distinctly perceive an object is referred to as the near point. At 10 to 30 years of age, the near point is about 9 cm in front of the eye. It gradually recedes from the eye, being about 28 cm at 45 years and 80 cm at 60 years. This condition is called presbyopia or old-sightedness. Presbyopia is related to a general condition called hyperopia or far-sightedness in which images are formed behind the retina.

Hyperopia, results when the length of eye is too short for the refractive power of the lens; this can result from having too short an eyeball or too weak a lens system. When a weak lens system results from age-related changes in the lens itself, the condition is presbyopia. The hyperopic eye sees distant objects clearly only by using the normal accommodative mechanism of the lens. It does this at the expense of constant contraction of the ciliary muscles, which produces eye strain.

Further, because part of the accommodative process is used to see even distant objects, less is available for viewing near objects; very near ones are never in focus. Because the image is focused behind the retina in hyperopia, the correction is accomplished by increasing the strength of the lens system, converging the light beam even more. This involves putting another convex lens in front of the eye. Another common deviation from the emmetropic, or normal eye, occurs when the image comes to focus in front of the retina.

This condition, called myopia or near-sightedness (because near objects are readily focused), results when the eyeball elongates too much for the refractive power of the lens, most commonly, myopia occurs beginning at puberty. For the myopic eye there are always points sufficiently near, from which light diverges enough to be brought to focus upon the retina, but light from all points at greater distances cannot be focused.

Ophthalmologists use the term far point to refer to the remotest point at which an object is clearly seen by the unaccommodated eye. In the myopic person, the far point is closer to the eye than normal. A concave (or negative, in terms of diopters) lens of the appropriate power (refractive index and curvature) will diverge light rays from more distant objects enough to compensate for the over-convergence of the eye and bring a focused image onto the retina of the lengthened eyeball.

Three other corrections for myopia have become popular. Radial keratotomy, making radial cuts in the cornea to flatten it out, can correct about 3 diopters of excess power, but the procedure has several disadvantages.

- It is difficult to accurately control the depth and length of the cuts, making the results somewhat unpredictable. In addition, too deep a cut can damage the endothelial layer on the inside of the cornea, the function of which is to pump water out of the cornea. This layer is essential to normal corneal function; humans cannot regenerate it.
- Epikeratophakos involves using the cornea from a donor eye as a living contact lens. The donor cornea is sewn to the recipient cornea. Some success has been achieved with this procedure.
- The third procedure involves fitting a special contact lens to the cornea. This lens "reshapes" the corneal surface to change its refractive power. Considerable success has been claimed for this procedure, but it is probably too new to know the permanence of the correction.

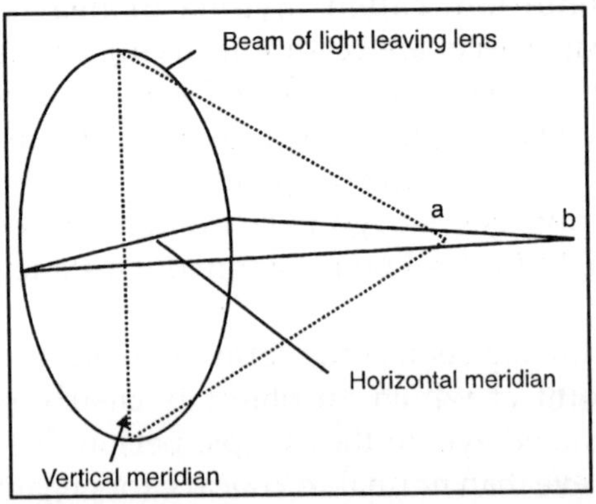

Fig. The Mechanism of Production of Astimatism.

Illustrated at left is a beam of light leaving the lens. Because the radius of curvature is greater along the horizontal meridian, the light passing through that meridian is focused at point b whereas light passing through the vertical meridian of lesser radius of curvature comes to focus nearer, at point a.

Yet another error of refraction, astigmatism results from asymmetry of the cornea. The ideal refractive surface is symmetrical in all meridians such that a line of light would come to focus at the same distance from the lens no matter how it was oriented. When this is not the case, an astigmatism results. That in the astigmatic eye a beam of light emerging from the lens along the vertical meridian is brought to focus at *a* and along the horizontal meridian (the axis of greatest radius of curvature) at point *b*.

If point *a* lies on the retina, then a line in the vertical meridian would appear focused; a line in the horizontal meridian would appear blurred. If point *b* were on the retina, the situation would be reversed. If the retina lies between *a* and *b*, then both would appear blurred. In the normal eye, points *a* and *b* would be the same. Astigmatism usually results from a cornea that is not spherical, but more egg-shaped with the long axis not parallel to the direction light is traveling to the retina.

Correction for astigmatism is accomplished by using a combination of spherical and cylindrical lenses, usually ground from a single piece of glass in such a way that refraction would be increased in the horizontal or decreased in the vertical meridian or both depending upon the location of the retina with respect to points *a* and *b*. An optometrist must determine how the asymmetry is oriented as well as how strong to make the corrective lens.

THE PUPIL

The functions of the pupil are to produce sharp images, especially in close work; to alter the depth of field; and to help regulate the amount of light striking the retina. When the pupil constricts, it prevents light from entering the periphery of the lens.

It is on the periphery where spherical aberrations, that is, unequal refraction of light by different parts of the lens, and chromatic aberrations, unequal refraction of light of different wavelengths, are greatest. Spherical aberrations occur because light passing through the edge of a lens is actually refracted more than light passing through its axis and comes to focus in front of the axial focus.

The image formed on the retina of a spot of light in the visual field is a spot surrounded by a blurred halo instead of a single spot. An entire object would appear blurred under these conditions. The pupil partially reduces the influence of this aberration when it constricts, allowing light to enter only through the centre of the lens. Light of long wavelengths, like red light, is refracted less than light of shorter wavelengths, like violet light, so the lens tends to separate wavelengths or colours, producing a rainbow effect. This effect (chromatic aberration) would be the worst on the periphery of the lens where spherical aberration is greatest.

Exclusion of the periphery of the lens is important in partially reducing the formation of coloured rings around an image. Constriction of the pupil or miosis also helps to reduce the rings around an image that result from light entering the eye from objects too far or too near to the lens to be in focus at a given radius of curvature. Controlling the depth of field, that is, the range of distances over which images are in focus for a given strength of the lens, is also a function of pupil diameter. When the eye is focused on objects at a great distance, the depth of field is normally large and pupil size is relatively unimportant.

When focused for near objects, however, the depth of field is small and can be increased markedly by closing down the pupil, just as it is by decreasing the aperture in a single lens reflex camera. The pupil also changes diameter with changes in light intensity at the retina, an increase causing the pupil to constrict and a decrease causing the pupil to dilate. This response is termed, the light reflex. Within limits, the light reflex provides an adequate amount of light on the retina, increasing the amount in dim light, decreasing it in bright light. The aperture of the pupil is not a constant function of light intensity because the sensitivity of the retina changes as a result of adaptation.

Thus, when the light intensity increases moderately, the pupil constricts, but then under constant illumination it gradually dilates again. With larger changes in intensity, the pupil will remain constricted. It is not unusual for the eye to experience a million-fold change in light intensity, but the pupil is only capable of about a four-fold change in diameter, that corresponds to about a 16-fold change in light intensity. This is not much compared to the total range the eye can experience; the greater part of the compensation is accomplished through retinal adaptation.

However, the pupillary reflex system is very fast compared to retinal adaptation that takes minutes. The light reflex is mediated by the autonomic nervous system. The afferent limb of the reflex involves the retina and the optic nerve. Fibres in the optic nerve give off collaterals to the pretectal nucleus, a structure just rostroventral to the superior colliculus.

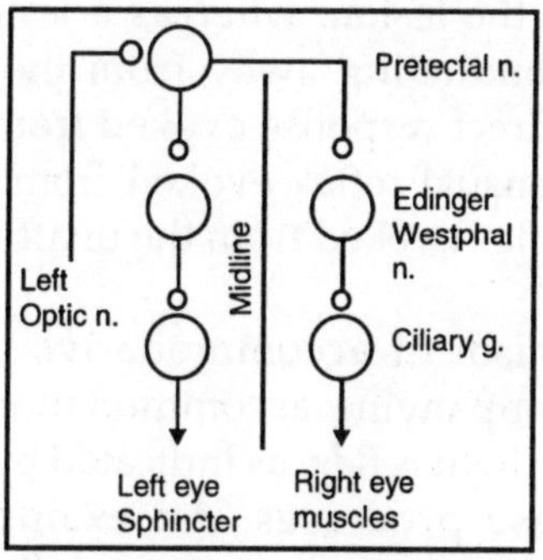

Fig. The Circuitry for Production of Both the Direct and Consensual Light Reflexes.

From this nucleus, fibres run bilaterally to the Edinger-Westphal nucleus, part of the oculomotor nucleus (third cranial nerve), forming connections with fibres of the third cranial nerve that are parasympathetic. These fibres relay again in the ciliary ganglion to innervate the sphincter pupillae. Increasing the amount of light at the retina increases the discharge to the various relays, finally causing contraction of the sphincter, closing down the pupil.

At the same time the sympathetic fibres that innervate the dilator pupillae are inhibited by this increased activity, relaxing this antagonistic muscle. Decreasing the amount of light at the retina causes an increase in sympathetic activity and dilation, while it reduces the output to the sphincter through the ciliary ganglion. The pretectal nucleus connects to the Edinger-Westphal nuclei bilaterally, which accounts for the fact that both pupils normally constrict together even though the light is directed into only one eye.

The constriction of the pupil through which the light is shown is called the direct light reflex, whereas constriction of the contralateral pupil is called the consensual light reflex. The existence of the consensual reflex is a great aid in determining the location of pathology involving this system.

A lesion of the afferent limb of the reflex (the part conducting toward the brain, including the optic nerve up to the pretectal nucleus) results in abolition of both the consensual and direct responses evoked by directing a light in the eye on the same side as the lesion, whereas a lesion of the efferent limb (the part conducting away from the brain) results in abolition of the direct response evoked from the affected side but not the consensual reflex evoked from that side (though the consensual reflex evoked from the unaffected side will also be abolished).

The mechanism of accommodative miosis (pupillary constriction accompanying accommodation) is not identical with miosis of the light reflex, as indicated by their dissociation in certain disease processes. An example is the Argyll Robertson pupil that occurs in most syphilitic infections of the central nervous system.

The pupil is constricted and is not responsive to light, but does respond normally during accommodation. The only regular pathological concomitant of syphilitic infections of the central nervous system is gliosis surrounding the aqueduct of Sylvius that has been interpreted to indicate damage to the Edinger-Westphal nucleus.

This implies that there are different cells for the two kinds of miosis, one type resistant to the spirochete. However, the cause of the Argyll Robertson pupil is not certain. In diphtheritic neuritis, the converse occurs: the pupil responds normally to light, but does not show accommodative miosis.

Dilation of the pupil is a normal concomitant of emotional states such as fright or alarm.

The dilator reflex is also elicited by most forms of noxious stimulation. Use has been made of the change in pupil size that occurs with pleasurable sensations in monitoring responses of patients in many clinical situations.

RECEPTIVE FUNCTIONS OF THE EYE

The photoreceptors of the eye are located within a thin sheet of tissue, the retina. Two types of receptors are present in the retina, the rods and the cones, so named because of their characteristic appearance. The human retina contains about 6.5×10^6 cones and 100×10^6 rods. Both rods and cones are found in the receptor layer of the retina, that outermost layer nearest the choroid. The receptors are connected to cells in the middle layer of the retina, the bipolar cells, which are in turn connected to cells in the innermost layer, the ganglion cells.

The axons of the ganglion cells are the fibres that eventually come together to form the optic nerve. Also in the retina are amacrine cells, that interconnect the bipolar cells with each other and the bipolar cells with ganglion cells, and horizontal cells, that interconnect receptor cells. These latter two cell types may play a role in sharpening images by surround inhibition, and the horizontal cells may play a role in colour opponent processes of colour vision. In addition, the retina contains neuroglia and blood vessels that support the tissue.

It is important to note that because of the way the retina is oriented, light entering the eye reaches the receptors only after passing through the network of blood vessels, cells, and fibres. The light finally reaches the light sensitive photopigment in the outer segments of the receptor cells, and what is not absorbed there is absorbed by the pigmented choroid, although a small part may be reflected back to the receptors. The receptors themselves consist of an inner and outer segment. The outer segment is composed of neatly folded lamellae that contain the photosensitive pigments that change their structure by isomerization when they absorb light.

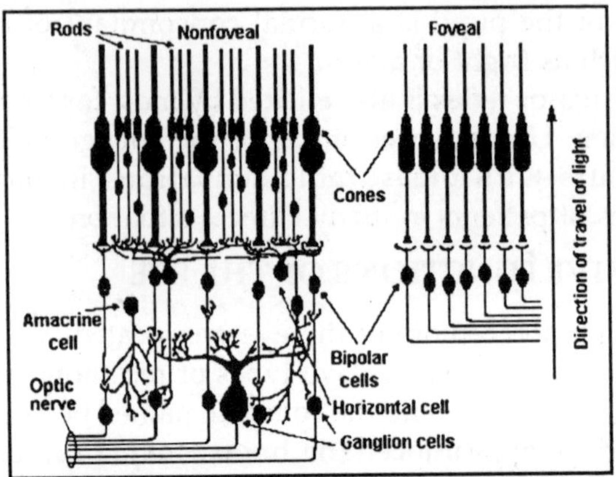

Fig. The Cellular Organization of the Foveal (Right) and Nonfoveal (Left) Retinas.

When they isomerize, they produce a change in the membrane potential of the receptor cell, the *receptor potential.* The term receptor potential is preferred here because no action potentials are generated in the receptors themselves. In fact, it is in the third-order ganglion cells that action potentials first appear in the visual system. The inner segment is joined to the outer segment by a narrow constriction that has characteristics of a true cilium. The inner segment contains mitochondria and endoplasmic reticulum and may be immediately adjoined to the cell body, as in cones, or connected to the cell body by a slender process, as in rods.

The short axon arises from the soma in both rods and cones. The retina may be partitioned into several regions that differ somewhat in structure. The central region or fovea ("pit") is a small depression of the retina (a concavity on the inner or vitreal side), where the retina is thinner than elsewhere. The fovea is about 1500 μm in diameter in man. Only cones are present in the fovea. The macula lutea or "yellow spot" is a region, including and surrounding the fovea, in which yellow pigment is found in the neuron layers of the retina.

In the region surrounding the fovea, there is a gradual decrease in the density of cones and a gradual increase in the density of rods until, at the outer margin of the retina, there are almost no cones at all. Finally, there is a region in which there are no receptors, but an accumulation of the ganglion cell axons to form the optic nerve where it leaves the eye. This region, called the optic disc, forms the physiological scotoma or blind spot.

RETINAL SENSITIVITY

Visual systems specialize in the detection of electromagnetic radiation. The human visual system is sensitive to radiation in wave lengths from 400-700 nm, that is, the spectrum of visible light from violet through red. Monochromatic sensitivity (sensitivity to one hue) varies continuously through the spectrum; thus thresholds are normally expressed as a curve called the visibility curve, relating sensitivity to wavelength. Actually, there are two such curves because the sensitivity varies with the immediate history of the eye.

These curves. If the eye has been in darkness for at least 30 minutes, and the tests are done with brief exposures, the curve on the left results. This is called the scotopic visibility curve, and it covers the range from 400 to 625 nm with a maximum sensitivity at about 507 nm, that is, sensitivity is maximum to blue-green light. If, on the other hand, the eye has been in normal daylight, the curve to the right, the photopic visibility curve, results. The maximum sensitivity for photopic vision is at 555 nm or in yellow-green light.

Notice that the values on the ordinate are normalized so that both curves reach a peak at a value of 1.0. If this were not the case, the scotopic curve would be much taller because the eye is much more sensitive after being in the dark than after being in the light. These properties are the result of the characteristics of the receptors that lie in the retina.

Rods and cones not only differ in their appearance, but also in their connections to the bipolar cells and in the kind of visual pigment they contain. It is these differences that account for many of the physiological properties of the retina. Cones in the fovea connect in a one-to-one fashion with bipolar cells that connect in a one-to-one fashion with ganglion cells. On the other hand, outside the fovea several rods and cones connect to a single bipolar cell.

Because many rods and cones connect to a single bipolar cell, a given amount of illumination will more strongly excite bipolar cells in the peripheral retina than in the foveal retina. In addition, the rods themselves are more sensitive to light, showing six times as much change in membrane potential per quantum of light as cones. Thus, the sensitivity of the retina is greater where there are more rods and less where there are only cones. This accounts for a phenomenon well known to sailors on watch.

A very faint light seen out of the corner of the eye, where the image falls on the peripheral retina, disappears when the eyes are turned toward the light, where the image falls on cones in the fovea. The greater sensitivity of the peripheral retina gives it superiority in scotopic or night vision, which lacks colour information, whereas the foveal retina normally functions in photopic or day vision, which is in colour. It is not clear why there is such an abrupt shift from rod to cone dominance in the photopic condition, but it appears that rod functioning is somehow suppressed when cones are active.

Table: Comparison of Peripheral and Foveal Retina

Property	*Foveal Retina*	*Peripheral Retina*
Threshold	Relatively high	Exquisitely low
Receptor distribution	Cones only	Rods and cones
Convergence	Limited or none in fovea; probably more elsewhere	Extensive
Illumination	Photopic (Daylight)	Scotopic (Night)
Functions	Central vision Detail vision Colour vision	Peripheral vision Achromatic vision Poor detail vision

ACUITY

The measure of acuity or detail vision is the minimum distance between two points, which are clearly seen as separate points, at a standard distance from the observer. Humans with normal acuity can distinguish two points separated by an angle of one minute of arc. This is usually expressed clinically as the reciprocal of the angle or 1.0; values less than one indicate reduced acuity.

Some people can distinguish separations as small as 0.5 to 0.33 minute of arc and therefore have acuities of 2.0 to 3.0. Because the retina is an array of receptors and not a continuous sheet, two point sources of light will only be seen as separate if the light from each one falls on a different receptor and the receptors are connected to separate communication lines to the brain, i.e., to separate ganglion cells. From what you have already learned about the retina, you should immediately conclude that acuity is greatest in the fovea.

Acuity is 1.0 in the fovea and falls to 0.1 within 30 degrees of the fovea. Acuity is, of course, zero in the blind spot where there are no receptors. The fovea is specialized for detail vision in several ways:

- The cones are more slender and more closely packed than elsewhere in the retina,
- Each cone has its own labeled line in the optic nerve,
- The blood vessels and extra cells are pushed out to the side of the fovea, reducing their light scattering to a minimum.

These factors combine so that the acuity near the edge of the macula is only half that at the fovea, and the acuity at the periphery of the retina is only one-fortieth that at the fovea. The fovea may also act as a lens to magnify the image in that region. Because the retina has a different (greater) refractive index than the vitreous, it could act as a diverging lens that would magnify the image on the dense array of cones.

ADAPTATION

Everyone has experienced the changes in visual sensitivity that occur when going from a situation of bright light into one of dim light, such as when entering a darkened theater. At first not much can be seen, but, after a period of time, a considerable sensitivity returns. This increase in sensitivity of the retina is called dark adaptation. The reverse condition, a super-sensitivity, followed by a decrease in sensitivity on emerging from the dark also occurs, and the decrease in sensitivity is similarly termed light adaptation.

The time course of dark adaptation (solid curve). In the graph, the log of the threshold for vision is plotted against time in the dark on the abscissa. Notice that the threshold decreases in two phases, one lasting about 7 minutes and the other beginning at about 7 minutes and completed by 30 minutes. During this time, the threshold has decreased by nearly one million-fold. The two phases probably represent changes first in the cones and then the rods. There are four pieces of evidence for this conclusion:

- If coloured light is used, the sensation is chromatic to the break point and then it becomes achromatic,
- If red light (rods are insensitive to red light) is used and directed only at the fovea, the second phase is absent,
- In persons with severe cases of congenital night-blindness (nyctalopia), in whom rod function is apparently completely absent, the second phase is absent,
- Individuals lacking cones, i.e., completely colour-blind persons, show only the second phase of adaptation, but starting at zero time in the dark.

Dark adaptation results from the fact that vision is a photochemical process. When light strikes the retina, some of the photons are absorbed by the molecules of the visual pigment embedded in the membranous disks of the receptors. When a molecule absorbs a photon it changes to a higher energy state. In this state, it may undergo a configurational change leading to decomposition of the pigment molecule.

The visual pigment of rods, rhodopsin, consists of a protein called opsin and an aldehyde of vitamin A_1 called retinene 1. When rhodopsin is decomposed to opsin and vitamin A_1, the conductance of the lamellated disk membrane (the membrane that contains the pigment) for calcium is thought to increase, allowing calcium to diffuse into the intracellular space, stabilize the outer receptor membrane and reduce the conductance for small ions, particularly sodium.

It is this reduction in the membrane conductance that is responsible for the receptor potential of the rods and cones. Reducing the sodium conductance results in an hyperpolarization of the membrane. In this decomposed or "bleached" condition, the pigment cannot absorb another photon and, therefore, cannot participate further in the visual process. However, there is a process, that involves active pumping of calcium into the membranous disks, that results in the reassembly of the pigment. The sensitivity of the receptor is a direct function of the amount of unbleached pigment it contains.

When moved into the light, pigment is bleached faster than it is reformed and there is a reduction in sensitivity. This is light adaptation. If enough pigment is bleached, a temporary blindness develops. On the other hand, when the light-adapted person goes into the dark, his sensitivity to light is low because most of his pigment is in the bleached condition, but because the pigment is being bleached at a lower rate, the regeneration process can reform the pigment faster than it is being bleached. The result is a gradually increasing quantity of unbleached pigment and thus a gradual increase in sensitivity or dark adaptation.

Under steady illumination, an equilibrium condition is reached between bleaching and reformation processes, and sensitivity of the retina stays constant. The exact position of this equilibrium depends upon the level of illumination. At first, it may seem puzzling that the rod adaptation curve is delayed. One reason for this is that cone pigments regenerate more rapidly than rod pigments. Also, the cone system seems to exert a suppressive control of the rod system. Clinically, this suppressive effect is manifest in the photophobia of the totally colour-blind person.

The totally colour-blind person functions well in dim light but has an abnormally (painfully) high sensitivity in bright light. This presumably reflects the lack of suppression of rods by the cone system. The break point in the adaptation curve of normal subjects comes at a later time in the adaptation period when a brighter light is used in the period before the dark adaptation is measured.

Again, the break points are indicated by arrows. This delay reflects an increase in light adaptation in brighter light that in turn reflects a greater amount of bleaching of the photosensitive pigments.

COLOUR VISION

It is the cone receptors that contain the visual pigments that are sensitive to the colour of light. In the 1800s, two physiologists, Thomas Young and Hermann von Helmholtz, proposed that there are three fundamental colour (hue) sensations in human vision, namely red, green, and violet (or blue), and that there are three different kinds of colour receptors, one for each of these colours.

Excitation of a given kind of receptor leads to the sensation of the appropriate colour, and all other colour sensations result from simultaneous excitation of more than one of these receptors. Thus, equal excitation of all three receptor types leads to the sensation of white light, whereas the appropriate amounts of excitation of red and green receptors yields the sensation of yellow, and so on. This theory, called the Young-Helmholtz, or the trichromatic theory, is the generally accepted theory today.

In the early 1960s, George Wald (a visual physiologist who later was awarded the Nobel prize) measured the sensitivity of the pigments of the fovea using psychophysical methods. Overall spectral sensitivity of the cones was measured by shining lights of various wavelengths on the fovea and determining the threshold strength for each. Then, the fovea was illuminated (adapted) with blue-green light (blue and green), presumably bleaching the green and blue pigments, leaving only the red functioning, and a new spectral sensitivity curve was measured.

Likewise, the fovea was adapted to purple (blue and red) light and the green pigment curve (curve G) was measured; the fovea was adapted to yellow (red and green) light and the blue pigment curve (curve B) was measured. The blue pigment curve shows peak sensitivity to light of wavelength about 440 nm, the green pigment to light of wavelength about 550 nm, and the red pigment to light of wavelength about 580 nm.

It should be noted that only the red pigment is sensitive at all to light of wavelength longer than 650 nm, though there is a rather broad overlap in sensitivity throughout the remainder of the visible spectrum. In addition, the sensitivity of the blue pigment is very much less than that of either of the others. Spectral sensitivities have also been measured for individual cones.

In this case, the absorption of light of various wavelengths by the pigment in a single cone is the variable measured instead of the sensitivity of the person as indicated by his verbal report, as used by Wald. The absorption of light of a given wavelength is determined by shining a light of that wavelength into the eye and measuring how much comes back out.

The difference between that amount and the amount that went in is the amount absorbed by the pigments. Plots of absorption spectra for eight cones, including two from human retinae and six from other primates.

There are clearly three different kinds of absorption spectra represented, with peak absorptions at about 445, 535, and 570 nm. These curves are not bad matches for those, considering that they were obtained by different methods and that one represents individual receptor responses, whereas the other represents a pigment population response. Evidence indicates that the three types of cones are not distributed uniformly across the cone portion of the retina. Blue cone density is greatest at the edge of the fovea (the human fovea is 1 degree in diameter) and falls off markedly by 10 degrees from the centre.

Red and green cones are distributed most densely in the fovea, with the density of green cones being greater than that of red cones. Over the entire retina the green cones are more dense than the red, which are in turn more dense than the blue. There are two ways of producing the various colours of the visible spectrum. One way is to shine a light of the appropriate wavelength in the eye.

The sensation of yellow light can be produced by shining a light of wavelength 590 nm into the eye. Alternatively, suitable amounts of light of three other appropriate wavelengths can produce the same sensation when shown into the eye at the same time.

The same sensation of yellow light can be produced by mixing light of wavelength 545 nm (green) and 670 nm (red). The observer looks through a hole in a partition at a white surface that is bisected by a black partition.

A yellow light of 589.3 nm is shown on the lower half of the field, and the observer reports seeing a yellow light. A light of 546 nm is shown on the upper half, and he reports seeing a green light; this is turned off and a light of 670.8 nm is shown on the upper half which he reports to be red. When both the red and green lights illuminate the upper field at the same time at the appropriate relative intensities, he reports that both the upper and lower fields are the *same* colour, namely yellow. This is the basis of function of the anomaloscope used to detect colour blindness.

Red, green, and blue were taken as the primary hues in this example, because these are the ones used by the human eye, but, from a physical point of view, any three other pure spectral hues could also be used. For the human eye, any colour experienced can be produced by a simple additive mixture of the correct proportions of red, green, and blue light.

COLOUR BLINDNESS

Colour blindness can be partially understood in terms of the mechanisms of colour vision. Essentially, there are three types of individuals with respect to colour vision, as follows:

Types of Colour Vision

- Trichromats
  - Normal
  - Protanomalous
  - Deuteranomalous

- Dichromats
  - Protanopes
  - Deuteranopes
  - Tritanopes
- Monochromats
  - Totally colour blind

There are persons, including those with normal colour vision, who have three visual pigments that they use in their colour perception. These are the trichromats. Those in a second class, the dichromats, match all the colours they see using only two colours; those in the third class, monochromats, see everything as a shade of gray, because they only see one colour, white. The protanomalous person has all three colour pigments, but he has a deficiency in his red responsiveness. In the anomaloscope, he requires at least 1.5 times more red intensity than normal in the upper field to match a yellow in the lower field.

The deuteranomalous person has reduced green sensitivity and requires at least four times more intense green than normal to match yellow. This is in fact how these abnormalities in colour vision are detected, by measuring the amounts, relative to the normal, of red and green that are required to match a standard yellow. As a consequence of this reduced sensitivity, a person with a colour anomaly can identify fewer hues than a normal person can. A normal person can distinguish about 150 different hues, whereas a protanomalous or deuteranomalous person distinguishes only 5 to 25 hues depending upon the severity of the deficiency.

In addition, there are some interesting contrast phenomena that show up in anomalous individuals. If a yellow light is shown in the lower field of an anomaloscope, the protanomalous individual identifies it as yellow, but if a green light is simultaneously shown in the upper field, he will report that the lower field is red. This does not happen in normal vision; the lower field remains yellow.

The dichromats present a different and more complicated picture. The luminosity curves for normal subjects, protanopes and deuteranopes. The protanope lacks red-sensitive pigment altogether. Notice the more rapid fall of the curve for protanopes at the long end of the spectrum (reds). This is the characteristic that distinguishes protanopes; they are completely insensitive to light beyond about 680 nm. The result is that they see little or no red.

They have been known to show up at a black-tie affair with a red tie on. The sensation of black is due to the absence of light and is therefore indistinguishable from red (of wavelength greater than 680 nm) by the protanope. The deuteranopes, on the other hand, are insensitive to green light. There are two types of deuteranopes.

There are those who have a reduced sensitivity to light in the middle and shorter wavelengths, which indicates an absence of green pigment, and there are those with normal spectral sensitivity whose problem seems to be that the red and green mechanisms are coupled together to form a single sensory system. This may occur in the retina at the level of the bipolar or ganglion cells, or it may occur more centrally in the visual system.

The tritanope is a rare individual who lacks sensitivity to the short end of the spectrum. Some people are of the opinion that tritanopia is not a retinal problem at all, but a disease of the optic nerve. Not enough information is available to say with confidence. No matter what his deficiency, the dichromat matches all colours with mixtures of only two pure spectral colours. The number of colours remaining chromatic varies with the type of dichromatism, but all of them are interpreted as various saturations of blue and yellow for the protanope and deuteranope, and red and green for the tritanope.

For each individual, there is a point of the visible spectrum between his two basic sensations that he matches with what a normal person sees as white. For the protanope this occurs at about 495 nm; for the deuteranope it occurs at 500 nm; for the tritanope it occurs at 572 nm. One unanswered question is why the protanope (and by analogy, the deuteranope) sees yellow at the middle (long end) of the spectrum and not green (red). It is possible that the pigment present (either red or green) excites both red and green mechanisms in the central visual system resulting in a sensation of yellow.

The cause of anomalous trichromacy (protanomaly or deuteranomaly) is usually stated to be an abnormally small amount of the appropriate pigment in the cones. This cannot be the case as evidenced by the experiment: A normal observer adjusts red and green intensities to match a yellow in an anomaloscope. He then looks away from the anomaloscope into an intense red light, bleaching the red pigment. If he looks immediately back into the anomaloscope, he will see green light in both the upper and lower fields, and both are the same green without adjustments of the intensity. If he adapts to green light instead, he will see the same red in both fields.

This indicates that the anomaloscope cannot distinguish the colour-adapted eye from the non-colour-adapted eye. We know that colour adaptation results from a simple reduction in non-bleached pigment. It follows that if the anomaly were due to such a simple reduction, the anomaloscope wouldn't detect it, but it does. A protanope can exactly match green with any red of wavelength less than 680 nm simply by adjusting the intensity of the red. As a result, an experimenter can exchange the matching red with the previously presented green, and the protanope will be none-the-wiser. The reason for this is that his green-cone pigment catches the light equally and experiences no change when they are switched.

The same trick cannot be played on a protanomalous individual; he sees the switch. His green pigment cannot detect it; this suggests he may have some anomalous pigment. In fact, such anomalous pigments have been identified by using the trick above to measure their sensitivity. The anomalies are probably due to a malfunction in the process for making the pigment. The correct red is replaced by an anomalous one that is less sensitive in the red region in the protanomalous individual; and, in the deuteranomalous individual, the green is replaced by anomalous pigment that is more like the red pigment.

The sum of these effects is that the remaining normal pigment (green in the case of protanomaly) and the anomalous pigment sensitivities are much closer together than the two normal pigments (red and green), and the individual has a harder time distinguishing colours in this region of the spectrum.

He thus requires more intensity of red to excite the anomalous pigment than he would to excite a normal red pigment to achieve the proper balance of green and red (anomalous red pigment) excitation. Protan and deuteran disturbances (terms for protanomaly and protanopia and for deuteranomaly and deuteranopia, respectively) are simple, sex-linked recessive traits. The trait is linked to the X chromosome, and so the condition is most common in males.

Females show the condition only if they have it present on both X chromosomes. These conditions occur in 1-13% of males, depending upon the country, but only 0.5% females. The normal condition is dominant to the protan or deuteran, and the anomalous condition appears to be dominant to the anopic. The control of green and red pigments seems to be at two separate loci on the chromosome, as indicated by family history data and the existence of women with normal colour vision, but with one deuteranope and one protanope son.

In the latter case, the women must have carried one defective gene of each type, yet neither was expressed in their phenotype. Deuteranomaly is the most frequently encountered defect, with protanomaly, deuteranopia, and protanopia less but about equally frequent.

Tritanopia only occurs in about 0.002% of the male population, and thus it is a rare event. The last colour vision defect is monochromatism. This usually shows up in the form of total colour blindness in which all colours are perceived as shades of gray.

People with this problem are photophobic, and show only scotopic luminosity curves. There are cases of subtotal monochromatism in which blue cones are not or only minimally involved or in which mainly macular cones are affected. The result is nearly the same because of the similarity of the blue cone and rod spectral sensitivities and because the macula is the major region used in vision in humans.

PHYSIOLOGY OF SINGLE VISUAL NEURONS

Visual Neurons in the Retina

The receptor cells and the bipolar cells of the retina respond to light with graded, electrotonic responses rather than all-or-nothing action potentials.

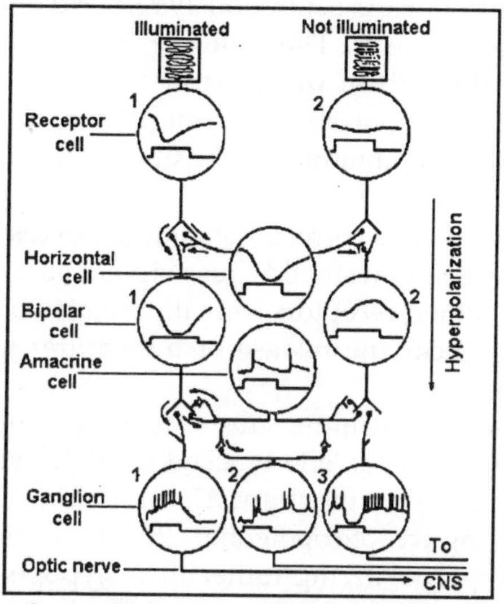

Fig. Synaptic Organization of the Vertebrate Retina and Responses of Retinal Neurons.

The graded responses in the receptors are the result of the photochemical process, but those in the bipolar cells are synaptically driven. Furthermore, it may be surprising that the receptors respond to light with an hyperpolarizing receptor potential, that is accompanied by an increase in membrane resistance.

A schematic diagram of a section of the retina with two receptors, one illuminated, the other unilluminated. The responses of the receptors, bipolar, ganglion, horizontal and amacrine cells are shown in circles representing each cell type. The time when a small spot of light was turned on is indicated by the upward deflection of the lower trace of each pair and the response of the cell by the upper trace. The hyperpolarizing responses of the illuminated receptor and its subjacent bipolar cell are illustrated as are the action potentials generated by the ganglion cell.

It is now known that in the dark there is a constant inward Na^+ current (dark current) flowing through the outer segment membrane and an outward current near the junction with the bipolar cell. This keeps the cell partly hypopolarized, and transmitter substance is continuously released onto the bipolar cell hypopolarizing it. The light flash decreases the dark current by the action of calcium to reduce membrane conductance, hyperpolarizes the cell relative to its dark state and decreases the amount of transmitter released onto the bipolar cell.

If this scenario is correct, then adding excess magnesium to the bathing solution should cause bipolar and ganglion cells to behave as if their receptors were illuminated because excess magnesium blocks the release of transmitter substances at chemical synapses.

A release of transmitter substance blocked by magnesium should be the same as one blocked by light, and it is. This scheme accounts mechanistically for the curious hyperpolarizing receptor potentials. But why are the receptor potentials hyperpolarizing rather than hypopolarizing as in other types of receptors?

It could be that we have identified the wrong stimulus! Perhaps it is darkness, not light, that is the stimulus. During evolution, when these receptors developed, the animals were living in the oceans and there were few land plants to cast shadows. Therefore, shadows cast by predators were relatively more frequent and predicted disaster.

If an animal could detect these shadows and respond quickly (by swimming away), he would have a survival advantage. Such a significant advantage would survive the evolutionary process.

The major line of transmission of information is from receptor to bipolar cell to ganglion cell and then to the brain, but the amacrine and horizontal cells provide lateral transmission lines that can produce the complicated centre-surround receptive fields of ganglion cells.

The horizontal cells receive their inputs from receptors as do the bipolar cells and, like the bipolar cells, they generate no spikes. Horizontal cell outputs inhibit (reduce transmission at) nearby unilluminated receptor-bipolar cell synaptic junctions. It has been suggested that they may enhance contrast by strongly turning off unstimulated bipolar cells. Amacrine cells produce action potentials and enter into reciprocal synaptic relations with bipolar cells.

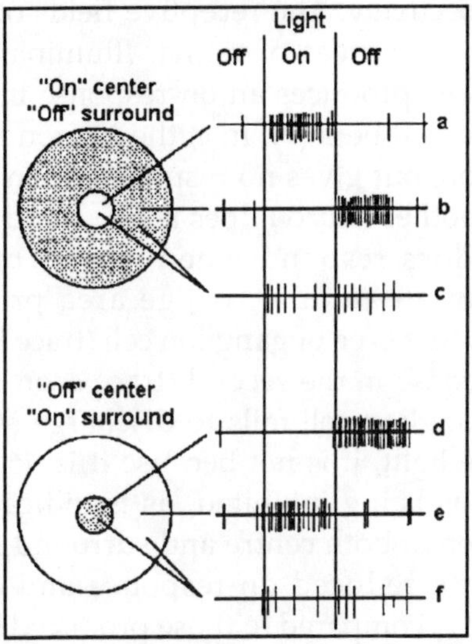

Fig. Receptive Fields of Two Retinal Ganglion Cells.

Fields are circuluar areas of the retina surrounded by an annulus of different properties. The cell in the upper part responds when the centre is illuminated (on-centre, a) and when the surround is darkened (off surround, b).

The cell in the lower part of the figure responds when the centre is darkened (off-centre, d) and when the surround is illuminated (on-surround, e). Both cells give on- and off-responses when both centre and surround are illuminated (c and f), but neither response is as strong as when only centre or surround is illuminated.

The response recorded from ganglion cells following light stimulation of the receptors is more conventional. Hypopolarizing synaptic potentials initiate trains of action potentials that propagate along the ganglion cell's axon. Ganglion cell responses are of three types:

- They respond (either discharge or increase their rate of discharge) only when the light is turned on,
- They respond only when the light is turned off;
- They discharge only at the beginning and end of a light period.

These are called on-responses, off-responses and on-off-responses, respectively. The receptive fields of ganglion cells are usually circular areas of retina. Illumination within the small central area produces an on-response in some cells (*a*), an off-response in others (*d*). In *a*, the neuron responds when the light goes on, but gives no response when it goes off (on-centre); in *d*, another neuron does not respond when the light goes on, but does respond when it goes off (off-centre). Illumination of an annular, fringe area produces an off-response in the first type of ganglion cell (trace *b*, off-surround) and an on-response in the second (trace *e*, on-surround).

When a ganglion cell fails to discharge at an on- or off-transient of the light, it is not because it is not being excited, but because it is being inhibited, as is indicated by the fact that illumination of both centre and surround simultaneously produces both a reduced on-response and a reduced off-response (*c* and *e*) compared to those produced by illuminating either centre or surround alone.

This effect is termed lateral inhibition or surround inhibition. The illuminated receptor cell 1 hyperpolarizes, causing synaptic transmission to be reduced to bipolar cell 1 and the horizontal cell. Transmission from bipolar cell 1 to ganglion cell 1 causes an hypopolarization that occurs when the light comes on (on-responses).

Transmitter substance is released by bipolar cell terminals in response to hyperpolarization; this is opposite to the situation at most chemical synapses. Receptor cell 1 is in the centre of the receptive field of ganglion cell 1, but it is in the surround of ganglion cell 3. The hyperpolarization of the horizontal cell by illumination of receptor cell 1 causes bipolar cell 2 to hypopolarize. This hypopolarization prevents the release of transmitter substance from bipolar cell 2 onto ganglion cell 3, and the ganglion cell does not discharge.

Actually, the ganglion cell is also inhibited (prevented from firing by hyperpolarization), perhaps through the amacrine cell. When the illumination of receptor cell 1 is turned off, bipolar cell 2 repolarizes.

This repolarization is equivalent as far as the transmitter release mechanism is concerned, to an hyperpolarization of the bipolar cell; it causes the release of transmitter substance onto the ganglion cell, which discharges (as the light goes off).

This mechanism would account for on-centre cells; presumably the mechanism for off-centre cells would be similar. Both rods and cones release the transmitter substance L-glutamate at their terminals on bipolar cells.

They do this in response to hyperpolarization when they are illuminated. The L-glutamate produces an hyperpolarization in "off" bipolar cells and an hypopolarization in "on" bipolar cells. Many people think that a given transmitter substance always produces the same effect in postsynaptic cells. In the receptor-bipolar cell synaptic arrangement, we see a clear indication that this idea is incorrect.

The L-glutamate activates a different kind of receptor in "off" and "on" bipolar cells. In "off" bipolar cells, activation of the KA/AMPA receptor by L-glutamate produces hyperpolarization. In "on" bipolar cells, the L-glutamate activates L-AP4 receptors to produce hypopolarization. The first stage of processing of colour information is apparently at the level of the ganglion cell, at least this is the first one studied to date. Colour-sensitive ganglion cells are sometimes called colour opponent cells, because cone mechanisms of different spectral sensitivities (i.e., red, green, and blue) interact in them antagonistically.

Most of these cells have concentric receptive fields like other ganglion cells, with either on- or off-centres, and the centres are sensitive to red, blue, or green. The annular surround of an on-centre gives an off-response for one of the other cone pigments. Cone mechanisms neighbouring each other in the spectrum appear to oppose each other so that red centres have only green surrounds and blue centres have only green surrounds, but green centres can have either red or blue surrounds. Red-blue interactions have not been found. Consider a red on-centre, green off-surround cell.

This cell discharges when the centre of the receptive field is illuminated with red light; it is inhibited when the centre is illuminated with green light, but discharges when a green light is extinguished after shining on the periphery; and it is inhibited by illuminating the periphery with red light.

The result of these interactions is a rather narrow spectral response of the central portion of the receptive field compared with the absorption curve for the pigments.

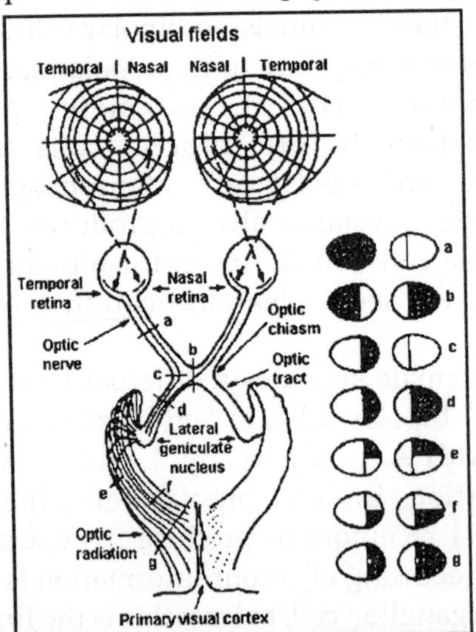

Fig. The Anatomic Organization of the Visual Pathway from the Retina to the Visual Cortex. Lesions of the Visual Pathway (a-g) Produce Defects in the Visual Fields as Indicated at the Right.

Thus, some of the very broad overlap in the absorption spectra of the neighbouring pigments is eliminated by the nervous system, which forms six fairly narrow spectral channels of information with both on and off information about the image.

Visual Neurons Outside the Retina

The axons of the ganglion cells form the optic nerves, which, after leaving the eyeball, proceed toward the brain until they come to the optic chiasm, where the optic nerves divide. Fibres from the nasal half of the retina cross to the opposite side of the brain; fibres from the temporal half go to the same side of the brain. Past the chiasm, crossed fibres from the contralateral eye join the uncrossed fibres from the ipsilateral eye to form the optic tract.

Fibres in the optic tract, which are still the axons of retinal ganglion cells, then proceed to the thalamus, where they end on cells of the lateral geniculate nucleus. The fibres of this nucleus project to neurons of the calcarine area of the occipital cortex.

This is area 17, as enumerated by the anatomist, Brodmann, and is often referred to as striate cortex because of its characteristic striped appearance under the microscope. There are also, direct projections to adjacent areas 18 and 19, often referred to as visual association areas. The receptive fields on the retina of single cells in the lateral geniculate nucleus are similar to those of ganglion cells.

Retinal receptive fields of geniculate cells have excitatory centres and inhibitory surrounds or vice versa, but the fields are larger, probably due to convergence of more than one ganglion cell on each geniculate neuron. Despite the presence of both crossed and uncrossed tract fibres in the geniculate, each cell there has a receptive field associated with only one eye. With respect to spectral sensitivity, geniculate neurons can be put into one of two categories:

- Narrow-band cells
- Broad-band cells.

Narrow-band cells respond with an increase in their discharge rate when their receptive fields are illuminated with light from part of the visible spectrum, and they decrease their discharge rate when their fields are illuminated with light from other parts of the spectrum. The majority of narrow-band cells either increase their response rates when their receptive fields are illuminated with red light and decrease their rates when illuminated with green light, or they increase their rates when illuminated with green light and decrease them when illuminated with red light.

The remainder of the cells increase their discharge rates when illuminated with yellow light and decrease them when illuminated with blue light or increase their rates for blue light and decrease them for yellow light. No red-blue cells have been found. In most respects, the narrow-band cells are not very different from colour-opponent ganglion cells except for the size of their receptive fields and the width of the spectrum to which they respond.

The peak sensitivity in terms of wavelength of the excitatory and inhibitory responses of geniculate cells corresponds nicely to the peak absorption spectra of the three cone pigments. Broad-band cells receive a uniform excitation or inhibition when their receptive fields are illuminated by light of all wavelengths and white light.

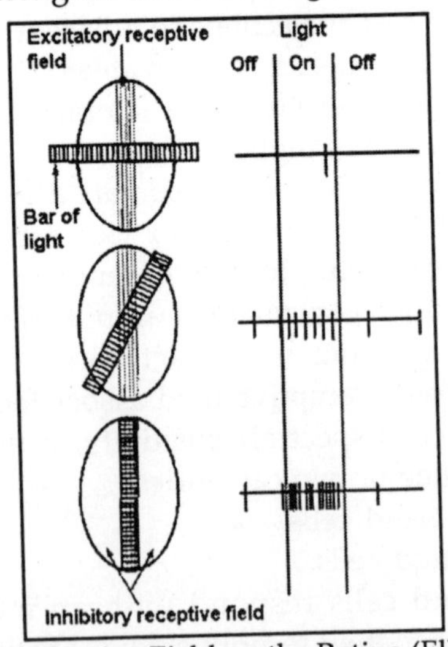

Fig. The Receptive Field on the Retina (Ellipse) of a Simple Cell in the Visual Cortex is Shown at the Left with a Bar of Light Superimposed on it at Various Angles.

The vertically striped portion of the receptive field is the excitatory area (excitatory receptive field), illumination of which excites the cell; the unhatched portion is the inhibitory area (inhibitory receptive field), illumination of which inhibits the cell.

The bar of light is striped across its length. The responses of the cell to the three different orientations of the light are shown at the right. The retinal receptive fields of neurons in the occipital cortex are complicated. Some visual cortical neurons (non-oriented neurons) possess receptive fields not particularly different from those of geniculate neurons; they have circular receptive fields and respond equally to stimuli of all orientations.

However, the receptive fields of most cortical neurons are not circular, most are arranged as parallel barlike excitatory and inhibitory areas with straight, rather than circular borders. Cortical cells with this characteristic receptive field are termed simple cells. In one sample of cortical neurons, two-thirds of the cells had simple receptive fields.

These cells can have two inhibitory regions flanking an excitatory region, the reverse situation can occur; or there can be just two regions side by side, one excitatory, the other inhibitory. As a result the cells respond best to narrow bars of light oriented in a particular direction across the retina. Rotation of the bar has two effects:

- It reduces the excitation, because less of the excitatory area is illuminated
- It increases the inhibition, because more of the inhibitory area is illuminated.

The result is that the cortical neuron responds less well when bars are not in their preferred orientation. Complex cells are similar to simple cells in that they respond to bars or lines of light (or the edges of bright objects) of proper orientation. An example of the response of a complex cell with a large receptive field, 8° x 16°, (indicated by the rectangle) to various illuminations.

At the right are shown records of discharges of a complex cell made with a patch of light off, then on, then off again, at the position indicated at the left of each record. The contour of the light patch is critical, because this cell responds to vertical lines. The cell responds vigourously when a light with a vertical edge is turned on (A) but hardly at all when the edge is sloping (B). The cell gives an on-response when the left part of the receptive field is illuminated (C and D) and an off-response when the right part is darkened (E and F).

However, it gives no response at all if the whole field is illuminated (G). The receptive fields on the retina of complex cells, in general, do not have characteristic excitatory and inhibitory areas, but they may respond with an on-response in part of the receptive field and an off-response in another part. Cells vary widely in which part does which. In general, complex cells seem to be responsive to lines of sharp contrast. However, unlike simple cells that show only on and off-responses, complex cells respond in a sustained manner to moving lines.

They respond better to a line moving in one direction than to the same line moving in the opposite direction across their receptive fields. The 2-sec records of the responses (on the right) of a complex cell to a bar of light (stippled rectangle) moving at various orientations with respect to the receptive field (3/8° × 3/8°, indicated by the broken square), as shown on the left.

The responses of the cell to movement, first in one direction and then the other, are indicated at the right. For this cell, the best response occurs when the long axis of the bar is parallel to the long axis of the receptive field and the bar is moving toward the right. Speed of movement of the bar also seems to be an important parameter for some cells.

A third class of visual cortical cells, the hypercomplex cell, is much like the complex cell in its response to lines of light or the edges of bright objects, but with the added complexity that it turns off if the line or edge is too long. This is clearly seen by reference which shows responses of a cell whose receptive field (2° × 2°, indicated by the broken square) was illuminated with a 1/8°-wide bar of light moving at 3°/sec. Up to a certain point, increasing the length of the line increases the response, increasing the length decreases the response.

In short, they behave as if the excitatory receptive field is bounded on the ends by an inhibitory field, but no effects (either excitatory or inhibitory) on cell discharge are seen by stimulating these boundary regions alone.

The right inhibitory flank is weaker than the left one. Many recent publications have dropped the term hypercomplex cells, and have substituted they term "complex cells with end-inhibition." The cells are still the same as before. Just how these cells are involved in creating vision is unknown, but some of the features that are abstracted are evident.

Contrast, lines, edges, movement, direction of movement, and length appear to be relevant variables in whatever these visual cells are doing. The colour sensitivity of visual cortical neurons is much the same as that of geniculate neurons. There are some with broad-band sensitivity and some with narrow-band sensitivity. Most of the narrow-band cells are either non-oriented or simple cells. They are excited by light of some wavelengths of the visible spectrum and inhibited by some other wavelengths, much as described earlier for geniculate cells.

Other cortical neurons, mainly some complex cells, are excited by illuminating their receptive fields with light from narrow bands at one end or the other of the spectrum. These narrow-band cells make up about 54% of all those cells in samples from the foveal representation in the striate (visual) cortex. The broad-band cells are very similar to broad-band cells in the lateral geniculate nucleus. We have as yet no idea how colour is perceived, but its perception seems to involve contrasting excitation and inhibition from the different types of cones or rather cone systems.

There is some indication that as a cell's colour specificity increases, its spatial selectivity decreases, implying that there are separate mechanisms (cells) for colour and form perception.

CLINICAL ASPECTS OF VISUAL FUNCTION

Light striking the nasal retina comes from the temporal visual field and that striking the temporal retina comes from the nasal visual field. The monocular visual field is the total extent of the visual world that is seen by one eye when it is not moving. It is customary to describe the defects of vision in terms of losses in the visual field, the measurement of which is called perimetry. The basic reference is a line perpendicular to a tangent to the centre of the surface of the cornea and passing through the centre of the fovea. This is the visual axis.

The visual axis is marked as the point in the centre of the concentric circles. The vertical diameter of the circles is the same for both eyes so that there is considerable overlap in the visual fields of the two eyes, the basis for stereoscopic vision.

The defects of the visual field that would develop from lesions at the levels indicated by the corresponding letters on the diagram to the left. It is possible to work out these defects from knowledge of the anatomy of the visual pathway. Points must be in mind:

- The retinal image is inverted top to bottom and left to right with respect to the visual field,
- Fibres from the nasal half of the retina cross over in the optic chiasm,
- Fibres from the upper part of the retina segregate from those from the lower,
- Upper retinal fibres from one eye stay with the upper retinal fibres from the other eye; the same is true for the lower retinal fibres,
- Fibres from the upper part of the retinae are found dorsally in the superior optic radiation passing through the parietal lobe, those from the lower part are ventrally in the inferior radiation (Meyer's loop) passing through the temporal lobe, and
- The macula may be represented on both sides of the cortex.

Loss of half of the visual field is called hemianopsia. If the visual field is half-lost on either the right or the left, in left optic tract section, the loss is called homonymous hemianopsia, in this case, right homonymous hemianopsia. If the loss is in both temporal fields, it is called bitemporal hemianopsia. Likewise, the nasal counterpart in binasal hemianopsia.

A quadrantanopsia results from interruption of either the superior or inferior optic radiation (clinically this usually involves a lesion in either parietal or temporal cortex) or alternatively from a partial lesion of the calcarine cortex, lower quadrants in the lingual gyrus, upper in the cuneate gyrus. The representation of the macula on both sides would result in macular sparing after a complete lesion of one visual cortex. In this condition, vision remains in the entire macula.

There are some investigators who believe that macular sparing is an artifact of inadequate control of eye movements during perimetry. As yet, this argument is unsettled. For a long time, it has been generally agreed that complete removal of the visual cortex in man leads to total, permanent blindness in the contralateral visual field, i.e., contralateral to the removed cortex. In the last ten years, it has become clear that this is not the case in the monkey. With total striate cortex removal, the monkey is still capable of locating briefly presented stimuli in space and can still discriminate visual patterns.

With less than total removal, the impairment is best characterized as a reduction in, not a loss of, sensitivity in part of the contralateral field, and, with training, the sensitivity gradually improves. When the lesion involves more than the striate cortex, especially when it includes the posterior parietal and temporal cortices, the animal is reduced to being able to tell only whether the light in its environment is on or off. In man, there is seldom a pure striate cortex lesion because the tissue is buried in the calcarine tissue. However, in a few cases of pure striate lesions, forced-choice discrimination methods have uncovered considerable visual function.

The methods involve instructing the patient to guess the identity or point to the location of a stimulus even though he does not see it. Remaining capacities are much like those in monkeys with similar lesions and include the ability to locate stimuli in space, to differentiate the orientation of lines (vertical versus horizontal versus diagonal, including white on black and black on white), to discriminate shapes (X versus O), and discriminate colours (red versus green).

The patients require moderately large stimuli; their performance falls off if the stimuli are too small. For the more difficult discriminations (where the stimuli are more alike, e.g., vertical versus diagonal lines), a slightly longer duration of exposure may be required.

Even though the human with the striate cortex lesion can perform without errors on such discriminations, he still reports that he saw nothing at all and expresses surprise when he is told of his performance.

These observations emphasize the fact that information about stimuli goes to more than one centre in the nervous system, and they show that what are sometimes called "subcortical relays" really do have functions. A person does not have to be able to report verbally the existence of a stimulus to respond appropriately to it. This sort of observation has led some investigators to parcel visual functions as:

- Recognizing, identifying, or examining functions, presumably the territory of cortical visual tissues.
- Detecting or orienting functions, presumably the territory of subcortical visual tissues. Only time will tell if this sort of distinction has meaning, but it does seem to fit the observations described.

There has been a flurry of activity designed to generate a prosthetic device for blindness. Such devices have been of two sorts:

The first represents an attempt to use some other modality with pattern recognition capabilities, and the second, more daring effort, actually attempts to produce a visual image. An example of the first type of device is a television camera whose output is transduced into an array of tactile stimuli that is applied to the skin of the back.

The camera is worn on the head, and the picture it sees is perceived by the wearer in sufficient detail that he can avoid obstacles while walking around. On the other hand, it can hardly be claimed that the person has vision.

The second type of device uses an array of photocells or a television camera whose output is applied to an array of small electrodes inserted into the retina or into the visual cortex itself. From what we know of the physiology and anatomy of the retina, it should come as no surprise that stimulation there has not reproduced visual experience. After all, the retina consists of about 106 million receptors, each independently sensitive to light. It seems unlikely that the activity of those receptors could be matched by a small array of electrical stimuli.

Electrical stimuli applied to the visual cortex typically produce sensations described as spots or balls of light that are either stationary or in motion.

In no case has the sensation been one of perceiving an object, even when an array of stimulating electrodes was used. Although all these devices have proven useful in helping blind people get around, none of them reproduces the lost vision.

Chapter 4

Photoreceptor of Sensory System

In a sensory system, a sensory receptor is a structure that recognizes a stimulus in the internal or external environment of an organism. In response to stimuli the sensory receptor initiates sensory transduction by creating graded potentials or action potentials in the same cell or in an adjacent one. The sensory receptor may be a specialized portion of the plasma membrane, a whole cell associated with a neuron ending, or a group of such cells.

PHOTORECEPTOR CELLS

The sensory receptors involved in taste and smell contain receptors that bind to specific chemicals. Odor receptors in olfactory receptor neurons, are activated by interacting with molecular structures on the odor molecule. Similarly, taste receptors (gustatory receptors) in taste buds interact with chemicals in food to produce an action potential. Other receptors such as mechanoreceptors and photoreceptors respond to physical stimuli.

Photoreceptor cells contain specialized proteins such as rhodopsin to transduce the physical energy in light into electrical signals. Some types of mechanoreceptors fire action potentials when their membranes are physically stretched.

The sensory receptor functions as the first component in a sensory system. Sensory receptors respond to specific stimulus modalities.

The stimulus modality to which a sensory receptor responds is determined by the sensory receptor's adequate stimulus.

The sensory receptor responds to its stimulus modality by initiating sensory transduction. This may be accomplished by a net shift in the initial states of a receptor.

PROPERTIES OF RECEPTOR

Let us begin our discussion of the neurophysiology of sensation by considering what happens in a receptor when a stimulus is applied to it. In later chapters individual sensory receptors for each sense will be considered separately, but at this point, it is the general properties of receptors that are of concern.

We have already seen that when a suprathreshold stimulus (of strength greater than threshold strength) is applied to a receptor, the nerve fibre associated with that receptor discharges.

RECEPTOR POTENTIALS

The transduction process appears to be similar in all of the various types of receptors that have been studied. As an example of a transducer, let us consider the muscle spindle, a small structure found in skeletal muscle that contains two kinds of receptors both of which can signal the length of the muscle. If a recording electrode is placed on the sensory nerve fibre supplying one of the receptors in the muscle spindle (the primary receptor) at a point very near the spindle itself and then the spindle is stretched.

The upper trace of each pair shows the recording from the fibre, whereas the lower trace shows the monitor of the muscle length (longer being indicated by an upward deflection). The spike discharge is superimposed upon a slow hypopolarizing potential that develops as the muscle is stretched. If the preparation is bathed in a solution of lidocaine, a local anesthetic agent, or in tetrodotoxin, the action potentials are blocked, leaving the slow hypopolarizing potential by itself.

This potential, called the generator potential, can be shown to originate in part of the spindle receptor, not in the nerve fibre.

There are receptors that are actually nerve fibres, part of which has been specialized to be sensitive to stimuli (e.g., receptors in the olfactory bulb and free nerve endings in skin), and there are receptors that are other types of cells in close association with nerve fibres (e.g., Merkel's disks in skin and sensory cells in taste buds). In the latter case, there appears to be a synaptic connection between the nonnerve cell and the nerve cell (the receptor), and the nerve fibres are usually not very sensitive to the same stimulus that excites their receptors. Some people use the term generator potential synonymously with receptor potential, whereas others prefer to reserve the term receptor potential for those cases where the receptor does not itself generate spikes, as in the receptors of the eye. The term receptor potential is used here in this latter sense.

The generator and receptor potentials are local or non-propagated events as shown by the degradation of their amplitudes with increasing distance from the receptor. One need only move the recording electrode along the nerve a few millimeters farther away from the position near the spindle that produced the records before the generator potentials disappear. They are graded responses as opposed to the all-or-none character of the action potential; the amplitude of the generator and receptor potentials increase with increasing stimulus strength.

Different amounts of muscle stretch, as shown by the heights of the indicators superimposed in the lower trace, resulted in the graded series of generator potentials superimposed. For these records, the action potentials have been blocked by application of lidocaine. The graded increases in the generator potential amplitude with graded stretches of the muscle are clearly illustrated.

The steps in formation of a generator potential are not known for every receptor, but where it has been studied the start of the generator potential usually results from an increase in the permeability of the membrane of the receptor to *all* small ions, but the ion furthest from its electrochemical equilibrium and in greatest concentration, namely sodium, contributes the greatest current.

As during an action potential, an increase in sodium conductance leads to an hypopolarization. The ionic mechanism of the generator potential is similar to that for the action potential but with a longer time constant; however, the restoration of the membrane potential to resting values at the end of the generator potential is a passive process not involving increased potassium conductance as in the action potential.

The transduction process takes place within the receptor region of the cell, that region that is sensitive to adequate stimuli and is responsible for the generation of the generator potential. The membrane of the receptor region is, however, electrically inexcitable; it contains no voltage-gated ionic channels and does not generate spikes. If the receptor region generated action potentials, the graded nature of the generator potential would be destroyed because as soon as the generator potential exceeded the critical firing level an action potential would be initiated, reversing the membrane polarization no matter how large or small the stimulus, i.e., the membrane potential would no longer encode the stimulus intensity.

The action potentials themselves are usually initiated in a physically separate region of the cell, the pacemaker or spike-generating region. This may be anatomically separate from or continuous with the receptor region or with the axon itself, the conducting region. For receptors that do not themselves generate spikes, the spike-generating regionis in a different cell. Receptor potentials occur in the rods and cones of the eye, but the first spikes in the visual system occur in the ganglion cells. That this must be true follows from Kirchhoff's current law; current flows only in complete circuits.

As anywhere in a nerve cell, outward current though an inactive (i.e., with normal resting conductances) region of membrane causes an hypopolarization of that membrane. An hypopolarization, as we already know, leads to the formation of action potentials, provided that the critical firing level is reached or exceeded. Actually, the spike-generating region simply responds to hypopolarizing current in the same way as any other membrane containing voltage-gated Na^+ and K^+ channels.

The conducting region of the membrane of the sensory receptor simply "follows" what the spike-generating region does; it simply transmits, without alteration, signals it receives. In the absence of an anesthetic agent, it is possible to study the relationship between the frequency of discharge in the spindle axon and the amplitude of the generator potential during the dynamic phase of stretch (i.e., the time when the muscle and the muscle spindle are actually changing length).

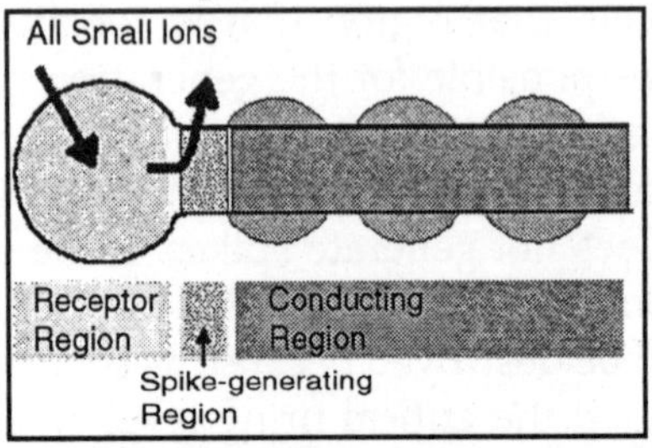

Fig. A Schemtatic Diagram:

A generaized receptor showing the receptor region, the spike generating region, and the conducting region along with the currents that flow during the generator potential.

The relationship obtained for various muscle lengths is linear. A linear relationship between the amplitude of the generator potential and the discharge frequency of the nerve has been found for every receptor studied so far. For the muscle spindle, there is also a linear relationship between the amount of stretch applied and the amplitude of the generator potential.

The muscle spindle is one of the cases of a receptor in which there is a linear relationship (the exponent of the power function, k=1) between the strength of the stimulus and the response of the cell, i.e., the amount of stretch and the frequency of discharge of the associated nerve fibre.

If the relationship between stimulus and response is curvilinear for a sensory receptor, as in those cases where the relationship is a logarithmic or power function (the exponent, k1), then there is a curvilinear relationship between the strength of the stimulus and the amplitude of the generator potential, and, in fact, it is a logarithmic or power function.

This must be the case if the relationship between the amplitude of the generator potential and the discharge frequency is always linear, because a linear transformation of any function or relationship always gives the same sort of function or relationship. It is important to bear in mind that the normal sequence of events in transduction is

- Stimulus
- Receptor or generator potential
- Action potentials

An adequate stimulus always leads to a receptor or generator potential if it is large enough and applied at the right place.

The generator potential normally leads to the formation of action potentials provided it is large enough to bring the membrane potential of the fibre to the critical firing level. In those cases where a receptor potential results from a stimulus, the receptor potential must somehow give rise to a generator potential before action potentials result.

In the retina of the eye, the receptor potential in the receptors, the rods and cones, causes a decreased release of a transmitter substance (a new stimulus, chemical this time) to the bipolar cells that causes them to generate a receptor potential.

The second receptor potential causes the release of a transmitter substance (a second new stimulus, again chemical) onto the ganglion cell that sets up a generator potential in the ganglion cell.

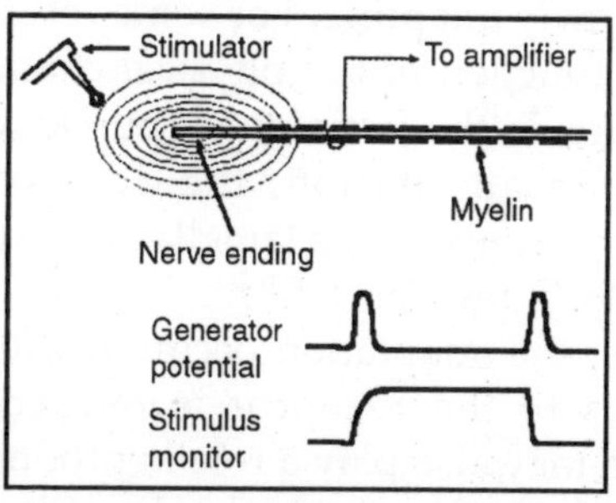

Fig. The Setup for Recording Generator Potentials from Pacinian Corpusles.

The generator potential in the ganglion cell finally initiates action potentials in the optic nerve fibres. In the retina, two receptor potentials and a generator potential intervene between the stimulus and the first action potentials, but the sequence of events is still: stimulus, receptor or generator potential, and action potentials.

ADAPTATION

The time over which summation can be obtained depends upon the duration of the individual generator potentials. To a certain extent, the duration of the generator potential depends upon the duration of the stimulus; however, some receptors have generator potentials that last only a short time, no matter how long the stimulus is maintained. The generator potentials of a Pacinian corpuscle and a muscle spindle obtained for a sustained deformation and stretch, respectively.

Notice that the generator potential for the corpuscle increases to a maximum amplitude and then returns to the resting membrane potential despite the persistence of the stimulus. On the other hand, the muscle spindle generator potential increases to a maximum and then declines a bit, but is maintained above the resting potential for the duration of the stimulus. We refer to a decrease in the amplitude of the generator potential or the frequency of discharge of the sensory fibre in the face of a persisting, constant stimulus as adaptation.

Receptors that behave like Pacinian corpuscles are said to be rapidly adapting; those that behave like muscle spindles are said to be slowly adapting. For some receptors, the rate of adaptation is a function of the physical characteristics of the receptor structure. In the Pacinian corpuscle, which shows an extreme of rapid adaptation, the adaptation is slowed considerably by removal of the lamellae and application of the stimulus directly to the nerve fibre.

In this case, the adaptation occurs because the coupling of the stimulus to the receptor is redueced. Membrane accommodation may also play a role in producing adaptation.

The reason for having both slowly and rapidly adapting receptors lies in the different kinds of information they signal.

Slowly adapting receptors presumably signal the onset and offset, but most importantly the presence of a long sustained stimulus.

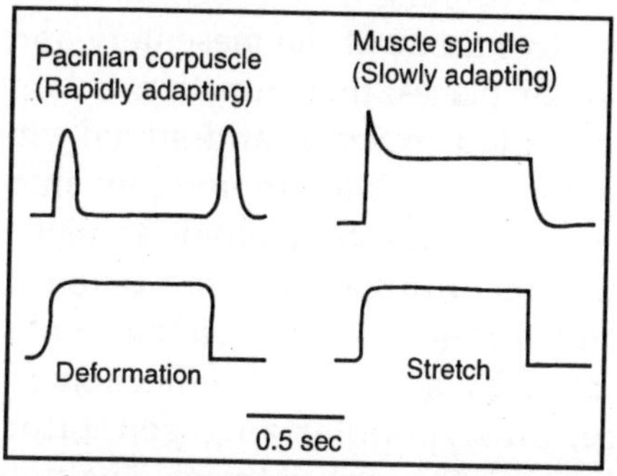

Fig. Comparison between Rapidly Adapting (Pacinian Corpuscle) and Slowly Adapting (Muscle Spindle) Generator Potentials.

They do adapt to some extent as indicated by the gradual waning of the sensation of the chair against your posterior as you read this. In contrast, rapidly adapting receptors signal the start, velocity, and, in some cases, the stop of a stimulus, but do not signal in between.

On the other hand, they give very good responses to repeated stimuli, suggesting they may play a role in sensations of vibration. The time over which summation of slowly adapting generator potentials can occur is longer than for rapidly adapting ones for a stimulus of the same duration.

SUMMATION

In receptors that are suitable for applying two stimuli close together, two other properties of generator potentials can be studied. If a mechanical stimulus is applied to a mechanoreceptor, a generator potential is recorded such that the amplitude of the generator potential is a function of the amount of deformation of the receptor surface. Two identical stimuli applied in rapid succession at the same site result in a generator potential twice the size of that resulting from a single stimulus.

In fact, the amplitude of the generator potential that results from any two stimuli applied in close temporal proximity is always equal to the algebraic sum of the generator potentials initiated by each stimulus alone. A tiny ellipsoidal body composed of a number of concentric lamellae and with a myelinated fibre running into its centre. The corpuscles are found in connective tissue of the mesentery and the popliteal fossa among other places; their function is largely unknown but they seem suited to sense and signal vibration. Their generator potentials have been thoroughly studied.

The way a mechanical stimulator is applied to the isolated corpuscle while recordings are being made from its axon. When the stimulator presses down on the corpuscle, as shown in the lower trace by an upward deflection of the stimulus monitor trace, an hypopolarizing generator potential is initiated in the receptor (upper trace). The potential rapidly returns to resting levels, but a new hypopolarizing generator potential is set up when the corpuscle is released from compression.

This behaviour is typical of the Pacinian corpuscle. If we were recording from the nerve fibre associated with the corpuscle at a site farther away, we would record an action potential associated with each of the generator potentials, but for now we are concerned about the generator potentials. We can use the first generator potential of the corpuscle to study summation. Stimulus A results in a generator potential of amplitude a and stimulus B (applied slightly later in time than A) in a potential of amplitude b.

When A and B are applied together, the resulting generator potential has amplitude $a + b$. This additive property of generator potentials is called temporal summation, because the variable is time. A similar phenomenon results when two stimuli are applied at the same time at two different locations. Stimulus C results in a generator potential of amplitude c, whereas stimulus A+C results in a generator potential of amplitude $a + c$. In this case, the variable is the spatial distribution of the stimuli instead of time; thus the phenomenon is termed spatial summation.

Summation of only two generator potentials, but any number of generator potentials can summate in a single cell, limited only by the physical arrangement of the receptor, i.e., how many stimuli can impinge upon it at the same time. The generator potential rises from the resting membrane potential, reaches a peak after a given interval, then returns toward the resting membrane potential. If the stimuli are presented so far apart in time that the first generator potential is over before the second stimulus is applied, then there will be no summation.

Thus, the amplitude of the summated generator potential depends upon how much the membrane potential of the receptor is altered from the resting potential at the moment the second stimulus is applied. The end result of the summation is an increase in the frequency of discharge of the sensory nerve fibre over the frequency that would be obtained for either stimulus alone. Any number of generator potentials can sum in this same manner, each increment in the amplitude resulting in increased discharge frequency, until a maximum frequency, characteristic of that particular fibre, is reached.

CODING

At one time the nervous system was thought to be a syncytium, a continuous, interconnected, multinucleated mass of protoplasm. The problem of explaining how information was routed from one place to another was most puzzling in such a system, because any stimulus should in theory be able to set the whole mass alive with impulses—all identical. How could any part of the system operate on its own; how could we tell hearing from smelling, high pitches from low?

We now know that nervous systems are composed of individual cells strung together and able to communicate with each other in specific patterns. Even with this "simplification," the problem of how the information contained within a stimulus is signaled to distant parts of the nervous system is formidable.

Any stimulus contains within it certain features that are of interest to the organism. Stimuli have

- Intensities or strengths
- Locations or sites of application
- Frequencies of application

- Rates of application
- Modalities

We begin our classification of stimulus features with modality, which, broadly speaking, is a class of sensations that are referred to a single type of receptor. Vision, hearing, touch, smell, and taste are all modalities. However, not every sensation can be associated with a single type of receptor, so the term modality becomes somewhat less precise. In the eye there are receptors for three different spectra of light and a fourth for black-and-white vision. Some physiologists prefer to define modalities of sensation in terms of the stimuli that produce them.

They refer to touch, pressure, and pain as modalities or submodalities of somesthesia with particular reference to the way the skin is stimulated. But, do touch, pressure, and pain represent different forms of stimulation or just different strengths of the same stimulus? With your fingers, lightly grip a fold of skin on your forearm, a stimulus you might consider as touch. A stronger grip yields a sensation of pressure, and an even stronger grip a sensation of pain. Yet the manner of stimulation is the same in all three cases.

Taking another example, electrical stimulation of a neuron can lead to the same sensation as a more natural stimulation of the skin. The term modality means one of the following:

- Vision
- Audition
- Gustation
- Olfaction
- Somesthesia (including skin, muscle, position, and visceral senses)

Any part of one of these modalities is called a submodality. Gustation, or the sense of taste, has only four submodalities—sweet, salty, sour, and bitter—whereas olfaction, the sense of smell, has a very large number. How do neurons recognize and encode the features of a stimulus? The term coding means simply the representation of the facts about the stimulus in terms of neural activity. Most neurons generate action potentials by which they communicate with each other.

These action potentials are essentially the same for all neurons. For the neurophysiologist studying coding, the problem is rather like that of a nation that intercepts a secret code sent by a foreign nation. There are quite a few codes possible using neural activity, many or all of which are used somewhere in the nervous system. For the present, the discussion will concentrate on five classes of neural codes, with some specific examples of how they may be used.

Neuron Pattern or Ensemble Codes

In some cases, one neuron may not be capable of signaling all the features of a stimulus that we can sense. In these cases, the stimulus features may be signaled by activity in populations of neurons and decoded by comparison, addition or subtraction of their activity. It is the ensemble of cells that carries the information about the particular sensation. An example of a possible pattern or ensemble code is the coding of the submodalities of taste: sweet, sour, salty, and bitter.

Every gustatory receptor responds to substances of all four qualities. Each column represents the responses of a single taste receptor to four different solutions, one from each submodality. The height of the bar indicates the magnitude of the response that particular receptor gave when the particular solution was poured over it. Looking at the response of only one receptor, it is impossible to tell what the stimulus was. Receptor 1 responds equally to salty and sweet and thus cannot distinguish between the two.

By comparing the responses of the four cells, it is possible to detect which stimulus was presented. Salty corresponds to large discharges from receptors 1 and 3, sweet to large discharges from receptors 1 and 4, and so on. There is a unique pattern of discharge across the group of receptors for each different quality, even though no one receptor can signal a given quality uniquely; obviously, there are more than four taste receptors in the tongue, but a similar principle could be applied to a much larger set of receptors.

Nonimpulse Codes

To this point, we have considered codes involving the generation of propagated action potentials. Why do nervous systems use action potentials to communicate information? Wouldn't it be easier to communicate increasing stimulus intensity with increasing amplitude of a voltage? The answer to both of these questions is quite simple. Voltages are attenuated drastically in a very short distance in volume conductors. Where long distances must be traversed it is imperative to use a brief, self-regenerating potential change that stays the same size all along the fibre. Coding can then be accomplished by modulating the frequency of discharge.

However, some cells in the nervous system do not generate action potentials at all. In these cases, the distances traversed are only micrometers, and reduction of potential with distance is no problem. In the eye, information must pass through a minimum of two cells before any action potentials are generated.

All of the information in the optic nerve discharges must also be encoded in the hypopolarizations and hyperpolarizations of the receptor cells, horizontal and amacrine cells, and bipolar cells.

Code of Specific Nerve Energies

Typically, sensory receptors are sensitive to many different kinds of stimulus energy. For instance, receptors in the hand respond to touch, heat, or vibration of the skin; the nerve itself can also respond to mechanical stimulation. Anyone who has ever hit his "funny bone" can attest to this fact. However, every receptor has one form of stimulus energy to which it is most sensitive, and this stimulus form is called the adequate stimulus.

The eye is excitable most easily by light, although it can also be excited by pressure on the eyeball itself; therefore, for the eye, light is the adequate stimulus. The neurophysiological term adequate stimulus, does not refer to stimulus strength, only to a form of energy. It is quite possible to have a subthreshold adequate stimulus.

Complementary to the principle of the adequate stimulus is a notion formulated by Johannes Müller. Müller noticed that, though the eye is excitable by light, pressure on the eyeball, electrical stimulation of the optic nerve, and some irritative pathological conditions, the sensation experienced is always one of vision—the person "sees light."

The Doctrine of Specific Nerve Energies, as formulated by Müller, says that, although a sense organ may be sensitive to many forms of stimulus energy other than its adequate stimulus, the sensation evoked is always like that associated with the adequate stimulus, no matter what kind of energy was applied. With electrical stimulation of the optic nerve, the sensation evoked is one of seeing light, not one of an electrical shock.

The doctrine of specific nerve energies implies that the modality or submodality of a sensation is determined not by the stimulus, but by what specific receptor or nerve fibre is stimulated. The doctrine also implies that the subjective qualities of a modality are determined, not in the receptors themselves, but in the central nervous system. An extension of the doctrine of specific nerve energies is the concept of labeled lines.

The concept of labeled lines says that information from a particular receptor travels over particular pathways to particular parts of the nervous system. Thus, the modality of a sensation depends upon which particular cell, pathway, nucleus or lobe is activated by the stimulus.

The concept of labeled lines receives some support from the fact that stimulation of the spinothalamic tract in man causes pain or from the observation that stimulation of the visual cortex leads only to a perception of light in a place within the visual world that depends upon where the visual cortex is stimulated. The concept of labeled lines implies that between the receptor on the periphery and the place in the central nervous system where the sensation occurs, there is no interaction between the modalities or submodalities.

This does not seem to be true. In the cuneate nucleus, a major relay in the somatosensory system, there is interaction between activity arising from hair receptors and activity arising from touch receptors.

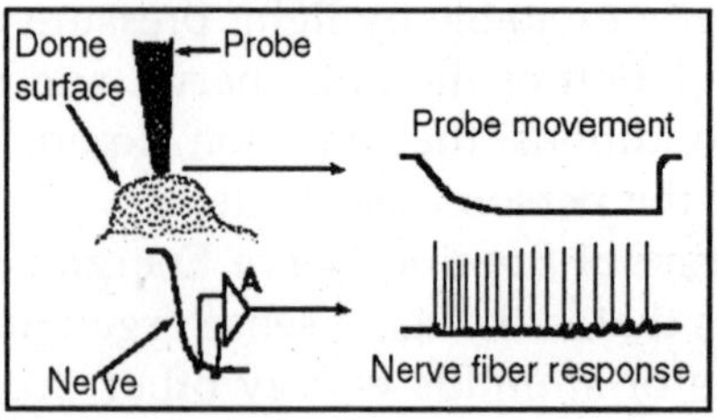

Fig. An Example of a Frequency Code.

In addition, there is interaction here between cutaneous, auditory, and visual activity. The nature of this interaction is just beginning to be explored.

It is true, however, that some structures within the central nervous system do appear to be associated with particular modalities of sensation. Even if a specific pathway or labeled line underlies the subjective quality of a sensation, a given neural pathway must use more than one code if other stimulus features are to be signaled. Whereas any activity in the pathway would evoke an experience of a specific quality or modality, the intensity of the sensation is usually coded within the activity itself.

Rate or Frequency Codes

Intensity can be coded in terms of discharge rate. Many mechanoreceptors use modulation of their discharge rates, that is, the number of impulses generated per second, to code the intensity of a mechanical stimulus. A rate code for the amount of displacement of the surface of a Merkel's disk mechanoreceptor found in the hairy skin of cats and on the abdomens of humans. In this experiment, a probe was applied to the dome-shaped surface of the receptor.

The probe was pushed down, indenting the surface a certain distance, and held there while recordings were made from the sensory nerve fibre supplying the receptor. The upper trace illustrates the movement of the probe (a downward deflection of the trace indicates a downward movement of the probe), whereas the lower trace is an actual recording of the fibre's discharge. The average frequency of discharge, measured during the time when the displacement was held constant (the plateau), for a number of different displacement distances are plotted.

Notice that the fibre codes increasing displacement by increasing its discharge rate in a linear fashion. The equation defining this relationship is F = 0.55D + 18, where F is the rate of discharge in spikes/sec and D is the displacement in micrometers.

Knowing this relationship, we can derive the amount of displacement the receptor must have experienced from the frequency of discharge of its nerve fibre.

INTERVAL CODE

Most neurons generate impulses or action potentials, but they spend most of their time in the resting condition. Even a neuron that is generating impulses at a rate of 50/sec spends most of its time resting.

If an action potential lasts 0.5 msec, as shown in A (some action potentials are shorter and some are longer), and repeats at a rate of 50/sec, as shown in B, the neuron is silent for 97.5% of the time.

Interestingly, one of the possible codes of neural activity is based on this silent time. For vibrating stimuli, it is possible to encode the frequency of the stimulus in a linear fashion with one impulse for each stimulus, as a frequency code, or with one impulse for every two or more vibrations of the stimulus, as an interval code. Coding of vibratory frequency is perhaps the most trivial example of a possible interval code and, at the same time, the least different from a frequency code.

When the stimulus is not periodic, the interval code is more obvious. It is possible that in some cells, such as pyramidal tract cells, a particular short interval between action potentials, in a train where the other intervals are much longer, may signal (code) the start of some movements. Briefly, however, this phenomenon is related to a special event called facilitation that occurs in certain postsynaptic cells only when two presynaptic action potentials arrive at short intervals.

TRANSDUCTION IN HAIR CELLS

Hair cells are found in the auditory and vestibular structures of the ear. They are columnar or flask-shaped and have an array of stereocilia at the apical ends. In the vestibular system there is one true cilium (the kinocilium) at one flank of the array of stereocilia.

The position of the kinocilium "polarizes" the cell. The arrangement of cilia and kinocilium in a hair cell in the vestibular system.

The kinocilia of the hair cells in the auditory system degenerate and so the cells are not effectively polarized. The cilia are bathed in potassium-rich endolymph, a situation different from what we normally encounter.

Normally, the extracellular fluid is poor in potassium; here it is rich. The hair cells are form synaptic connections from neurons at the basal ends (the end opposite the stereocilia); these connections can be either afferent (sensory) or efferent (activity coming from the CNS, presumably to exert control of sensory functions).

According to this model, a stimulus that displaces the stereocilia toward the kinocilium (or in the direction where the kinocilium would be in the auditory system), elongates the tip link that pulls on the mechanical gate and opens it. The channel then admits potassium ions; their entry hypopolarizes the hair cell.

A schematic diagram of the way this is thought to occur. Conversely, when the stereocilia are displaced away from the kinocilium, the tension on the tip link is reduced (perhaps it shortens) reducing the pull on the mechanical gate and closing it. Potassium no longer enters the cell, and it repolarizes.

The cation-selective channels in hair cells are blocked by aminoglycoside antibiotics, e.g., streptomycin. Higher doses have toxic effects on hair cells, damaging the stereocilial bundles and eventually killing the cells. With protracted hair cell bundle displacement, the amplitude of the generator potential gradually decreases, i.e., there is adaptation. The rate and extent of the adaptation is known to increase with increasing calcium concentration in the endolymph.

This process is thought to occur by the mechanism. It appears that the structure anchoring the tip like at its upper end actually moves toward the base of the stereocilium. Current thought is that myosin, the same substance involved in contraction of skeletal muscle, is activated by calcium ions. The calcium ions enter the cell through the same channel that admits the potassium ions that hypopolarize the cell.

Calcium, through its interaction with calmodulin and thence with myosin, causes the channel to migrate toward the base of the hair cell, thereby reducing the tension on the tip link and allowing the channel to close.

Closing the channel reduces the inward current and allows the cell to repolarize, i.e., to adapt. For hair receptors the transduction involves a conversion of mechanical energy to the electrical energy of spikes.

In the case of auditory receptors, the mechanical energy occurs in the form of alternating compressions and rarefactions of some medium, usually air.

In the case of vestibular receptors, the mechanical energy occurs in the form of displacements of hairs due to the force of gravity or to inertia.

OLFACTORY RECEPTORS

Things that we smell are varied. Attempts to reduce the number of smelled qualities to a few categories have been largely unsuccessful in providing only a few categories. All things smelled are aerosolizable and water soluble because they must reach and dissolve in the mucus that covers the olfactory receptors.

Unlike the hair cell receptors, the olfactory receptors are themselves sensory neurons. They have cilia that project into the mucus surrounding the olfactory epithelium.

It is on these cilia that the odorant receptors are thought to be located. Each olfactory neuron expresses only one type of odorant receptor.

Binding of the odorants to receptors activates a G protein, which in turn activates adenlyl cyclase to produce cAMP. It is cAMP that activates a cyclic nucleotide-gated cationic channel.

When the channel is opened, sodium and calcium enter the cell, hypopolarizing it.

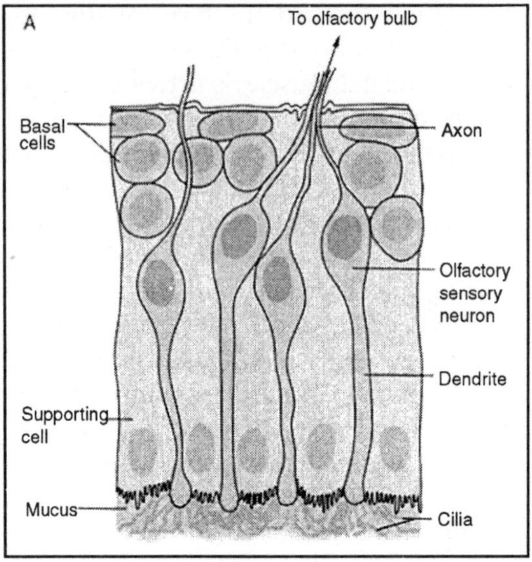

Fig. Four Olfactory Sen-sory Neurons are Shown with their Cilia Projecting into the Mucus.

The location of the receptors and second messenger cascades and the sequence of events in channel opening.

MUSCLE SPINDLE RECEPTORS

The muscle spindle lies within skeletal muscles. It is arranged in parallel with the extrafusal fibres of the muscle, those that actually do the shortening. In this position, the receptors are optimally placed to sense the length of the muscle and the rate of change of length. The receptor cells themselves (themselves sensory neurons) are actually associated with small striated muscle fibres within the spindle.

On the surface of the sensory neurons are located the stimulus-gated channels. These channels are interconnected with each other and other elements of the cytoskeleton by cytoskeletal strands of spectrin. The channels in question are opened by stretching the membrane; probably the spectrin strands "pull" the channel open when the membrane is stretched.

So, mechanical deformation of the membrane opens these cation-selective channels, and there is an influx of sodium and /or calcium, which hypopolarizes the membrane (the generator potential). If large enough this hypopolarization will lead to spikes. The presumed physical arrangement of the channels and spectrin strands. It is not difficult to see how this arrangement might lead to receptor activation and spike generation.

The muscle spindle receptors are used by the brain to determine the lengths of the various skeletal muscles and to derive from the lengths the angles of the joints. Clearly, as a muscle changes length, the tension in the muscle changes, and as the tension in the muscle increases the receptor structures. In this case, the transduction is a change from mechanical energy to the electrical energy of spikes.

The graded quality of the receptor potential is clearly seen on the left. On the right are shown traces of membrane currents made by "patch-clamp" recording. Horizontal bars are placed over the periods of time when the channel is open. A downward deflection here shows an inward current. Note that the channels open only briefly, then close. Activation of the channel causes it to open more frequently and to stay open longer.

VISUAL RECEPTORS

Visual receptors must convert the energy in light to action potentials, at least indirectly. The visual receptors, the rods and cones, do not support action potentials. Their membrane potentials do change in response to light stimulation, but their membranes contain no voltage-gated channels and therefore they cannot produce spikes.

There are two kinds of visual receptors–the rods and the cones, so named because of their characteristic shapes. Both types of receptors have an inner segment, consisting of the cell body and efferent process, and an outer segment, connected to the inner segment by a cilium and consisting mainly of layers of folded membrane. The folded membrane of the external segments contains the visual pigment, a substance that absorbs photons and changes its conformation. All the rods contain the pigment, rhodopsin. The cones contain one of three different pigments.

The outer segment membrane contains cGMP-gated sodium channels, which are open in the dark. That means that there is a continuous entry of sodium ions into the cell in the dark, the dark current, that hypopolarizes the membrane of the receptor. Stimulation of the receptors by light turns off the dark current allowing the membrane to repolarize, usually referred to as an hyperpolarization (relative to the state that exists in the dark). The cascade of events that leads from the photon to the cessation of the dark current is thought to involve a G protein. Specifically, the conformational change in the photopigment activates a G protein, which leads to activation of a cGMP phosphodiesterase.

The resulting conversion of cGMP to 5'-GMP reduces the amount of cGMP in the cell. Sodium channels gated by cGMP will then close, reducing the inward. We have seen that generator or receptor potentials in all types of receptors except those in the eye are hypopolarizing.

A generator potential must hypopolarize because it is the hypopolarization that leads to the action potential. Why then are receptor potentials in the eye of an hyperpolarizing nature? Why are they different?

Perhaps it is because these potentials are not generator potentials–they don't lead directly to action potentials. However, consider the possibility that we have misidentified the stimulus for the eye.

We like to think of the stimulus as light, the lack of a stimulus as dark. It is possible that the real stimulus is dark. Consider the plight of some ancient ancestors of vertebrates, the ostracoderms. These jawless fish-like creatures probably swam around their environments with their mouths open, filtering the water near the floor of lakes or rivers for edible morsels.

Clearly the ostracoderms were not predators. Now, the ostracoderms were probably a favourite food of an invertebrate predator, the eurypterids or water scorpions.

Clearly, the eurypterids overmatched the ostracoderms based on their much larger size. The only defence for the ostracoderm was its bony outer covering, hence the name. Speculate with me for a moment. Suppose that these mainly defenceless fish-like creatures developed a type of cell (a receptor) that contained a photopigment and was capable of secreting a substance (let's call it a hormone) that could cause the animal to begin swimming.

Further suppose that the substance was released from the cells when the receptor was shaded–it's a shadow detector. During the Ordovician period when these animals lived there were few large land plants to cast shadows across the water. Because of this, the major source of shadows would have been other creatures, among them the eurypterids.

An approaching predator would cast a shadow upon the primitive photoreceptor, the cell would release its hormone and the ostracoderm would swim away to safety. Clearly, this would convey a survival advantage on the animal who possessed such an ability.

Now suppose that an ostracoderm possessed such a photoreceptor but with a longish process that brought the release point for the hormone closer to the target muscles. That animal would have an advantage over one without such a process. According to Darwinian logic, that animal would be selected for survival. The longer the process, the more likely the survival. Gradually, the cellular arrangement would come to look less like an element of an hormonal system and more like an element of a nervous system.

If this scenario has any accuracy at all, then the first photoreceptors may well have been shadow detectors for which the stimulus was a shadow, i.e., darkness. By this logic, then our visual receptors may well have developed from such shadow detectors.

During the process of evolution, we came to use these receptors in a different way, emphasizing the presence of light more than its absence. In light of this argument, it is clear that the actual stimulus for the photoreceptor, in paleontologic terms, may not be light but dark.

GUSTATORY RECEPTORS

Gustatory receptors, located on the tongue and nearby, are specialized to detect things that are sweet, sour, salty or bitter. There are many other qualities of foods that we associate verbally with taste; all but these four are actually associated with olfactory receptors. Taste buds (groups of gustatory receptors) are comprised of receptors arranged like segments of an orange.

Each receptor has microvilli on its apical end. The microvilli project through the surface of the tongue in taste pores. Like the hair-cell receptors, the gustatory receptors are not themselves nerve cells, but they receive innervation from nerve cells near their bases, again much like hair cells. The arrangement of receptors and nerve cells in a taste bud.

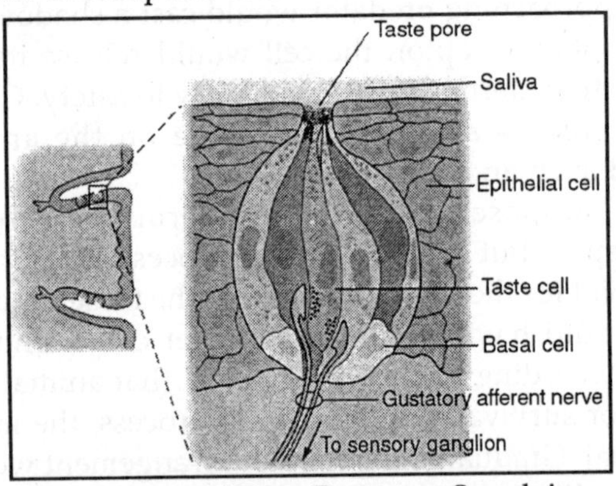

Fig. A Group of Taste Receptors Supplying their Microvilli to the Taste Pore.

Tastants (things we taste) produce receptor potentials in different ways.

- Some tastants simply enter the receptor cell through channels as ions. Salty substances often contain sodium ions. These sodium ions can simply enter the receptor cell through sodium or cationic channels. Sour substances are acidic. The hydrogen ions can enter cells through cationic channels.
- Other tastants can compete for use of potassium channels, thereby reducing outward potassium

currents (this would result in hypopolarization). Bitter substances like quinine can block the potassium channels, leading to hypopolarization.

- Still other tastants work through second messengers to close potassium channels, reducing potassium current. Both bitter and sweet substances act in this way.
- A final group of tastants act through second messengers to open chloride or non-specific ion channels.

Chapter 5

Medical Imaging

Medical imaging refers to the techniques and processes used to create images of the human body for clinical purposes (medical procedures seeking to reveal, diagnose or examine disease) or medical science (including the study of normal anatomy and function).

As a discipline and in its widest sense, it is part of biological imaging and incorporates radiology (in the wider sense), radiological sciences, endoscopy, (medical) thermography, medical photography and microscopy (e.g. for human pathological investigations). Measurement and recording techniques which are not primarily designed to produce images, such as electroencephalography (EEG) and magnetoencephalography (MEG) and others, but which produce data susceptible to be represented as maps, can be seen as forms of medical imaging.

In the clinical context, medical imaging is generally equated to Radiology or "clinical imaging" and the medical practitioner responsible for interpreting (and sometimes acquiring) the images is a radiologist.

Diagnostic radiography designates the technical aspects of medical imaging and in particular the acquisition of medical images. The *radiographer* or *radiologic technologist* is usually responsible for acquiring medical images of diagnostic quality, although some radiological interventions are performed by radiologists.

As a field of scientific investigation, medical imaging constitutes a sub-discipline of biomedical engineering, medical physics or medicine depending on the context: Research and development in the area of instrumentation, image acquisition (e.g. radiography), modelling and quantification are usually the preserve of biomedical engineering, medical physics and computer science;

Research into the application and interpretation of medical images is usually the preserve of radiology and the medical sub-discipline relevant to medical condition or area of medical science (neuroscience, cardiology, psychiatry, psychology, etc) under investigation. Many of the techniques developed for medical imaging also have scientific and industrial applications.

Medical imaging is often perceived to designate the set of techniques that noninvasively produce images of the internal aspect of the body. In this restricted sense, medical imaging can be seen as the solution of mathematical inverse problems.

This means that cause (the properties of living tissue) is inferred from effect (the observed signal). In the case of ultrasonography the probe consists of ultrasonic pressure waves and echoes inside the tissue show the internal structure. In the case of projection radiography, the probe is X-ray radiation which is absorbed at different rates in different tissue types such as bone, muscle and fat.

MODERN IMAGING TECHNOLOGY

Fluoroscopy

Fluoroscopy produces real-time images of internal structures of the body in a similar fashion to radiography, but employs a constant input of x rays. Contrast media, such as barium, iodine, and air are used to visualize internal organs as they work. Fluoroscopy is also used in image-guided procedures when constant feedback during a procedure is required.

Magnetic Resonance Imaging (MRI)

A Magnetic Resonance Imaging instrument (MRI scanner) uses powerful magnets to polarise and excite hydrogen nuclei (single proton) in water molecules in human tissue, producing a detectable signal which is spatially encoded resulting in images of the body. In brief, MRI involves the use of three kinds of electromagnetic field:

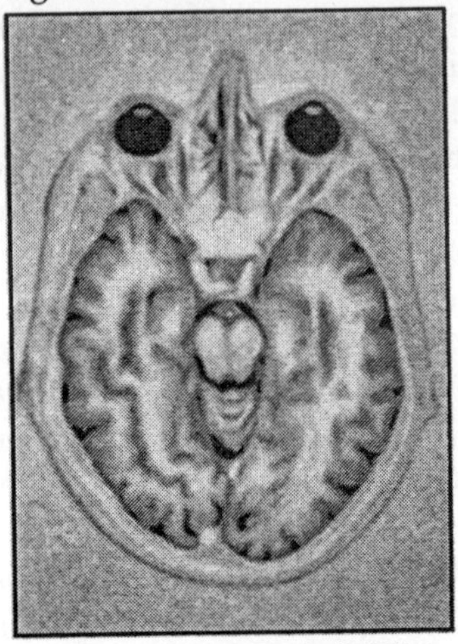

Fig. Human Brain MRI

A very strong static magnetic field to polarize the hydrogen nuclei, called the static field; a weaker time-varying (of the order of 1 kHz) for spatial encoding, called the gradient field(s); and a weak radio-frequency (RF) field for manipulation of the hydrogen nuclei to produce measurable signals, collected through an RF antenna.

Like CT, MRI traditionally creates a 2D image of a thin "slice" of the body and is therefore considered a tomographic imaging technique. Modern MRI instruments are capable of producing images in the form of 3D blocks, which may be considered a generalisation of the single-slice, tomographic, concept. Unlike CT, MRI does not involve the use of ionizing radiation and is therefore not associated with the same health hazards; there are no known long term effects of exposure to strong static fields and therefore there is no limit on the number of scans to which an individual can be subjected, in contrast with X-ray and CT.

However, there are well identified health risks associated with tissue heating from exposure to the RF field and the presence of implanted devices in the body, such as pace makers. These risks are strictly controlled as part of the design of the instrument and the scanning protocols used. CT and MRI being sensitive to different properties of the tissue, the appearance of the images obtained with the two techniques differ markedly.

In CT, X-rays must be blocked by some form of dense tissue to create an image, therefore the image quality when looking at soft tissues will be poor. While any nucleus with a net nuclear spin can be used, the proton of the hydrogen atom remains the most widely used, especially in the clinical setting, since it is so ubiquitous and returns much signal. This nucleus, present in water molecules, allows excellent soft-tissue contrast.

MRI, or "NMR imaging" as it was originally known, has only been in use since the early 1980s. Effects from long term, or repeated exposure, to the intense static magnetic field are not known.

Nuclear Medicine

Images from gamma cameras are used in Nuclear Medicine to detect regions of biological activity that are often associated with diseases. A short lived isotope, such as I is administered to the patient. These isotopes are more readily absorbed by biologically active regions of the body, such as tumors or fracture points in bones.

Positron Emission Tomography (PET)

Positron emission tomography is primarily used to detect diseases of the brain and heart. Similarly to nuclear medicine, a short-lived isotope, such as ^{18}F, is incorporated into a substance used by the body such as glucose which is absorbed by the tumor of interest.

PET scans are often viewed along side computed tomography scans, which can be performed on the same equipment without moving the patient. This allows the tumors detected by the PET scan to be viewed next to the rest of the patient's anatomy detected by the CT scan.

Projection Radiography

Radiographs, more commonly known as x-rays, are often used to determine the type and extent of a fracture as well as for detecting pathological changes in the lungs. With the use of radio-opaque contrast media, such as barium, they can also be used to visualize the structure of the stomach and intestines - this can help diagnose ulcers or certain types of colon cancer.

Photoacoustic Imaging

Photoacoustic imaging is a recently developed hybrid biomedical imaging modality based on the photoacoustic effect. It combines the advantages of optical absorption contrast with ultrasonic spatial resolution for deep imaging in (optical) diffusive or quasi-diffusive regime. Recent studies have shown that photoacoustic imaging can be used in vivo for tumor angiogenesis monitoring, blood oxygenation mapping, functional brain imaging, and skin melanoma detection etc.

TOMOGRAPHY

Tomography is the method of imaging a single plane, or slice, of an object resulting in a tomogram. There are several forms of tomography:

Linear Tomography

This is the most basic form of tomography. The X-ray tube moved from point "A" to point "B" above the patient, while the cassette holder (or "bucky") moves simultaneously under the patient from point "B" to point "A." The fulcrum, or pivot point, is set to the area of interest. In this manner, the points above and below the focal plane are blurred out, just as the background is blurred when panning a camera during exposure. No longer carried out and replaced by computed tomography.

Poly Tomography

This was a complex form of tomography. With this technique, a number of geometrical movements were programmed, such as hypocycloidic, circular, and elliptical. Philips Medical Systems produced one such device called the 'Polytome.' No longer carried out, replaced by computed tomography.

Zonography

This is a variant of linear tomography, where a limited arc of movement is used. It is still used in some centres for visualising the kidney during an intravenous urogram (IVU)

Orthopantomography

The only common tomographic examination in use. This makes use of a complex movement to allow the radiographic examination of the mandible, as if it were a flat bone. It is often referred to as a "Panaray", but this is incorrect, as it is a trademark of a specific company's equipment

Computed Tomography (CT)

A CT scan, also known as a CAT scan, is a helical tomography (latest generation), which traditionally produces a 2D image of the structures in a thin section of the body. It uses X-rays.

It has a greater ionizing radiation dose burden than projection radiography; repeated scans should be limited.

CLINICAL IMAGING OR BIOLOGICAL IMAGING TECHNIQUES

Electron Microscopy

The electron microscope is a microscope that can magnify very small details with high resolving power due to the use of electrons as the source of illumination, magnifying at levels up to 2,000,000 times.

Electron microscopy is employed in anatomic pathology to identify organelles within the cells. Its usefulness has been greatly reduced by immunhistochemistry but it is still irreplaceable for the diagnosis of kidney disease, identification of immotile cilia syndrome and many other tasks

Creation of Three-dimensional Images

Recently, techniques have been developed to enable CT, MRI and ultrasound scanning software to produce 3D images for the physician.

Traditionally CT and MRI scans produced 2D static output on film. To produce 3D images, many scans are made, then combined by computers to produce a 3D model, which can then be manipulated by the physician. 3D ultrasounds are produced using a somewhat similar technique.

With the ability to visualize important structures in great detail, 3D visualization methods are a valuable resource for the diagnosis and surgical treatment of many pathologies. It was a key resource for the famous, but ultimately unsuccessful attempt by Singaporean surgeons to separate Iranian twins Ladan and Laleh Bijani in 2003. The 3D equipment was used previously for similar operations with great success.

Other proposed or developed techniques include:

- Diffuse optical tomography
- Elastography
- Electrical impedance tomography
- Optoacoustic imaging
- Ophthalmology
  - A-scan
  - B-scan
  - Corneal topography
  - Heidelberg retinal tomography
  - Optical coherence tomography
  - Scanning laser ophthalmoscopy

Some of these techniques are still at a research stage and not yet used in clinical routines.

Non-diagnostic imaging: Neuroimaging has also been used in experimental circumstances to allow people (especially disabled persons) to control outside devices, acting as a brain computer interface.

Medical imaging service: This is a specialized area of medical equipment service and repair, which is separate from the biomedical field, although a hospital with their own service group may include them in the biomed department.

At one time, there were only two ways to receive training for this field. One was to learn it in the military, and the other was on-the-job training (OJT) from the manufacturer. But since the 1980s several independent training centres have been started. One such school is RSTI.

There are different means of employment in this occupation. Working for the manufacturer's field service department (OEM), working for a hospital (in-house), and working for an independent (outside, or independent provider). The most stable positions are with the OEM or hospital, as you can remain current through on-going training, and the two have good working relationships.

The OEM service engineer can expect to spend a lot of time driving from one site to another during the work day, and working non-standard hours. They will install, remove, diagnose, repair, calibrate, perform preventive maintenance, and interface equipment, all while ensuring good customer relations.

The in-house person is employed by the hospital. With larger medical facilities, travel between the hospitals' other locations may be required to perform the required services. You may also be required to do yearly testing of the radiation sources for compliance. The OEM or independent will provide installation of purchased equipment, and can be used for back-up service.

An independent is typically someone who has left an OEM, and started their own service business. Staying up-to-date as an independent can be difficult and expensive, as the OEM is usually reluctant to provide training. However, non-OEM training facilities are available, such as the aforementioned RSTI. Competition for service can be aggressive, with OEM's giving hospitals or clinics a reduction in equipment purchase price if they retain some form of OEM service.

The independent may also sell and install refurbished equipment, or de-install equipment. They will repair, calibrate, and perform preventative maintenance. Because many of the tasks associated with imaging service require expensive, specialized equipment, there may be a financial limit to the independent. Typical equipment used routinely are a Storage Oscilloscope and multimeter (if servicing old vacuum-state equipment, a VOM would be helpful). Additional equipment: Keithley dosimeter, mAs meter, Biddle contact tachometer, light to radiation template, etc.

MAGNETIC RESONANCE IMAGING

Magnetic resonance image showing a median sagittal cross section through a human head.

Magnetic resonance imaging (MRI) is primarily used in medical imaging to visualize the structure and function of the body. It provides detailed images of the body in any plane. MR has much greater soft tissue contrast than Computed tomography (CT) making it especially useful in neurological, musculoskeletal, cardiovascular and oncolological diseases. Unlike CT it uses no ionizing radiation.

The scanner creates a powerful magnetic field which aligns the magnetization of hydrogen atoms in the body. Radio waves are used to alter the alignment of this magnetization. This causes the hydrogen atoms to emit a weak radio signal which is amplified by the scanner. This signal can be manipulated by additional magnetic fields to build up enough information to reconstruct an image of the body.

Magnetic resonance spectroscopy is used to measure the levels of different metabolites in body tissues. The MR signal produces spectrum of difference resonances that correspond to different molecular arrangements of the isotope being "excited". This signature is used to diagnose certain metabolic disorders, especially those affecting the brain, as well as to provide information on tumor metabolism.

The scanners used in medicine have a typical magnetic field strength of 0.2 to 3 teslas. Construction costs approximately US$ 1 million per tesla and maintenance an additional several hundred thousand dollars per year. Research using MRI scanners operating at ultra high field strength (up to 21.1 tesla) can produce images of the mouse brain with a resolution of 18 micrometres.

Background

Nomenclature

Magnetic resonance imaging was developed from knowledge gained in the study of nuclear magnetic resonance. In its early years MRI was referred to as nuclear magnetic resonance imaging (NMRI), but the word nuclear has been associated with ionizing radiation exposure, which is not used in an MRI, so to prevent patients from making a negative association between MRI and ionizing radiation, the word has been almost universally removed. Scientists still use the term *NMRI* when discussing non-medical devices operating on the same principles.

One of the inventors of MRI, Paul Lauterbur, originally named the technique *zeugmatography*, a Greek term meaning "that which is used for joining". The term referred to the interaction between the static and the gradient magnetic fields necessary to create an image, but the nomenclature never caught on.

Brief Explanation for the Layperson

MRI works by making certain atoms in the body emit radiowaves by using a powerful magnet. When a person is in the scanner some of the hydrogen atoms, which are mainly found in water, align with this magnetic field. A radiowave at just the right frequency makes these atoms resonate and absorb energy.

The atoms then release this energy in the form of a very weak radiowave that can be measured by the scanner and amplified. Extra magnetic fields are used to constantly change the magnetic field to allow images of the body to be reconstructed.

These fields are created by gradient coils which make the familiar banging sounds of an MRI scan. Contrast agents may be injected to demonstrate blood vessels or inflammation in the tissues.

Unlike CT scanning MRI uses no ionizing radiation and is generally a very safe procedure. Patients with some metal implants and cardiac pacemakers are prevented from having an MRI due to effect of the powerful magnetic field.

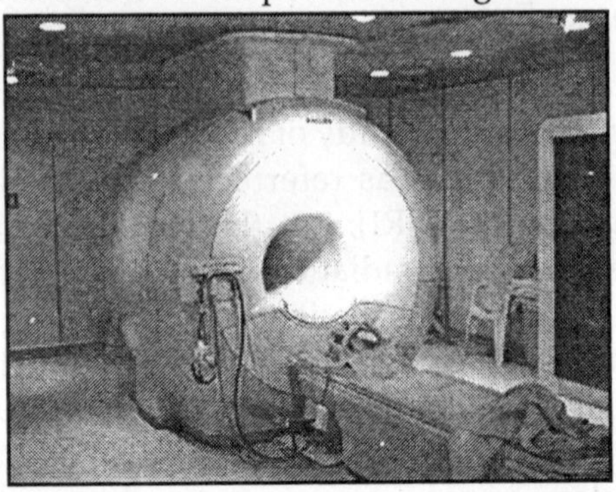

Fig. Modern 3 Tesla Clinical MRI Scanner.

MRI is used to image every part of the body, but is particularly useful in neurological conditions, disorders of the muscles and joints, for evaluating tumors and showing abnormalities in the heart and blood vessels.

PRINCIPLE

Magnetism

Elementary subatomic particles such as protons have the quantum mechanical property of spin. Nuclei such as ^{1}H or ^{31}P, with an odd number of nucleons, always have a non–zero spin and therefore a magnetic moment. Some other isotopes such as ^{12}C have no unpaired neutrons or protons, and no net spin.

When these spins are placed in an external magnetic field they start to precess around the direction of that field. The magnetic field also creates two energy states that the protons can occupy which are separated by a quantum of energy. The thermal energy of the sample causes the molecules to tumble leaving only a very small excess of protons to cause magnetic polarization.

Resonance

The energy difference between the proton energy states corresponds to electromagnetic radiation at radio frequency wavelengths. Resonant absorption of energy by the protons due to an external oscillating magnetic field (radio wave) will occur at the Larmor frequency.

The net magnetization vector has two components. The longitudinal magnetization is due to an excess of protons in the lower energy state. This gives a net polarization parallel to the external field. The transverse magnetization is due to coherences forming between the two proton energy states. This gives a net polarization perpendicular to external field in the transverse plane. The recovery of longitudinal magnetization is called T1 relaxation and the loss of phase coherence in the transverse plane is called T2 relaxation.

When the radio frequency pulse is turned off, the transverse vector component produces an oscillating magnetic field which induces a small current in the receiver coil. This free induction decay (FID) lasts only a few milliseconds before the thermal equilibrium of the spins is restored. The actual signal that is measured by the scanner is formed by a refocusing gradient or radio wave to create a gradient or spin-echo.

Imaging

Slice selection is achieved by applying a magnetic gradient in addition to the external magnetic field during the radio frequency pulse. Only one plane within the object will have protons that are on–resonance and contribute to the signal.

A real image can be considered as being composed of a number of spatial frequencies at different orientations. A two–dimensional Fourier transformation of a real image will express these waves as a matrix of spatial frequencies known as k–space. Low spatial frequencies are represented at the centre of k–space and high spatial frequencies at the periphery. Frequency and phase encoding are used to measure the amplitudes of a range of spatial frequencies within the object being imaged.

The frequency encoding gradient is applied during readout of the signal and is orthogonal to the slice selection gradient. During application of the gradient the frequency differences in the readout direction progressively change. At the midpoint of the readout these differences are small and the low spatial frequencies in the image are sampled filling the centre of k-space. Higher spatial frequencies will be sampled towards the beginning and end of the readout filling the periphery of k-space.

Phase encoding is applied in the remaining orthogonal plane and uses the same principle of sampling the object for different spatial frequencies. However, it is applied for a brief period before the readout and the strength of the gradient is changed incrementally between each radio frequency pulse. For each phase encoding step a line of k–space is filled.

Magnet

The magnet is the largest and most expensive component of the scanner, and the remainder of the scanner is built around it. Just as important as the strength of the main magnet is its precision. The straightness of magnet lines within the centre or, as it is known as, the iso-centre of the magnet, needs to be nearly perfect. This is known as homogeneity. Fluctuations (non-homogeneities in the field strength) within the scan region should be less than three parts per million (3 ppm). Three types of magnet have been used:

- *Permanent magnet*: Conventional magnets made from ferromagnetic materials (e.g., steel) can be used to provide the static magnetic field. These are extremely bulky (the magnet can weigh in excess of 100 tonnes), but once installed require little costly maintenance. Permanent magnets can only achieve limited field strength (usually < 0.4 T) and have limited stability and precision. There are also potential safety issues, as the magnetic field cannot be removed in case of entrapment.
- *Resistive electromagnet*: A solenoid wound from copper wire is an alternative to a permanent magnet. The advantages are low cost, but field strength is limited, and stability is poor. The electromagnet requires considerable electrical energy during operation which

can make it expensive to operate. This design is essentially obsolete.

- *Superconducting electromagnet*: When a niobium-titanium alloy is cooled by liquid helium at 4 K ("269 °C, "452 °F) it becomes superconducting where it loses all resistance to flow of electrical current. By building an electromagnet from superconducting wire, it is possible to develop extremely high field strengths, with very high stability. The construction of such magnets is extremely costly, and the cryogenic helium is expensive and difficult to handle. However, despite its cost, helium cooled superconducting magnets are the most common type found in MRI scanners today.

Most superconducting magnets have their coils of superconductive wire immersed in liquid helium, inside a vessel called a cryostat.

Despite thermal insulation, ambient heat causes the helium to slowly boil off. Such magnets, therefore, require regular topping-up with helium.

Generally a cryocooler, also known as a coldhead is used to recondense some helium vapour back into the liquid helium bath. Several manufacturers now offer 'cryogenless' scanners, where instead of being immersed in liquid helium the magnet wire is cooled directly by a cryocooler.

Magnets are available in a variety of shapes. However, permanent magnets are most frequently 'C' shaped, and superconducting magnets most frequently cylindrical. However, C-shaped superconducting magnets and box-shaped permanent magnets have also been used.

Magnetic field strength is an important factor determining image quality. Higher magnetic fields increase signal-to-noise ratio, permitting higher resolution or faster scanning. However, higher field strengths require more costly magnets with higher maintenance costs, and have increased safety concerns. 1.0 - 1.5 T field strengths are a good compromise between cost and performance for general medical use. However, for certain specialist uses (e.g., brain imaging), field strengths up to 3.0 T may be desirable.

RADIO FREQUENCY SYSTEM

The radio frequency (RF) transmission system consists of a RF synthesizer, power amplifier and transmitting coil. This is usually built into the body of the scanner. The power of the transmitter is variable, but high-end scanners may have a peak output power of up to 35 kW, and be capable of sustaining average power of 1 kW.

The receiver consists of the coil, pre-amplifier and signal processing system. While it is possible to scan using the integrated coil for transmitting and receiving, if a small region is being imaged then better image quality is obtained by using a close-fitting smaller coil. A variety of coils are available which fit around parts of the body, e.g., the head, knee, wrist, or internally, e.g., the rectum.

A recent development in MRI technology has been the development of sophisticated multi-element phased array coils which are capable of acquiring multiple channels of data in parallel. This 'parallel imaging' technique uses unique acquisition schemes that allow for accelerated imaging, by replacing some of the spatial coding originating from the magnetic gradients with the spatial sensitivity of the different coil elements.

However the increased acceleration also reduces the signal-to-noise ratio and can create residual artifacts in the image reconstruction. Two frequently used parallel acquisition and reconstruction schemes are SENSE and GRAPPA. A detailed review of parallel imaging techniques can be found here:

Gradients

Magnetic gradients are generated by three orthogonal coils, oriented in the *x*, *y* and *z* directions of the scanner. These are usually resistive electromagnets powered by sophisticated amplifiers which permit rapid and precise adjustments to their field strength and direction. Typical gradient systems are capable of producing gradients from 20 mT/m to 100 mT/m (i.e. in a 1.5 T magnet, when a maximal *z*-axis gradient is applied the field strength may be 1.45 T at one end of a 1 m long bore, and 1.55 T at the other). It is the magnetic gradients that determine the plane of imaging - because the orthogonal gradients can be combined freely, any plane can be selected for imaging.

Scan speed is dependent on performance of the gradient system. Stronger gradients allow for faster imaging, or for higher resolution; similarly, gradients systems capable of faster switching can also permit faster scanning. However, gradient performance is limited by safety concerns over nerve stimulation.

In order to understand MRI contrast, it is important to have some understanding of the time constants involved in relaxation processes that establish equilibrium following RF excitation. As the high-energy nuclei relax and realign they emit energy at rates which are recorded to provide information about the material they are in.

The realignment of nuclear spins with the magnetic field is termed *longitudinal relaxation* and the time required for a certain percentage of the tissue's nuclei to realign is termed "Time 1" or T1, which is typically about 1 second at 1.5 Tesla main field strength. T2-weighted imaging relies upon local dephasing of spins following the application of the transverse energy pulse; the *transverse* relaxation time is termed "Time 2" or T2, typically < 100 ms for tissue at 1.5 Tesla main field strength.

A subtle but important variant of the T2 technique is called T2* imaging. T2 imaging employs a spin echo technique, in which spins are refocused to compensate for local magnetic field inhomogeneities. T2 imaging is performed without refocusing.

This sacrifices some image integrity (resolution) but provides additional sensitivity to relaxation processes that cause incoherence of transverse magnetization. Applications of T2* imaging include functional MRI (fMRI) or evaluation of baseline vascular perfusion (e.g. cerebral blood flow (CBF)) and cerebral blood volume (CBV) using injected agents; in these cases, there is an inherent trade-off between image quality and detection sensitivity. Because T2*-weighted sequences are sensitive to magnetic inhomogeneity (as can be caused by deposition of iron-containing blood-degradation products), such sequences are utilized to detect subtle areas of recent or chronic intra cranial hemorrhage.

Image contrast is created by using a selection of image acquisition parameters that weights signal by T1, T2 or T2, or no relaxation time ("proton-density images"). In the brain, T1-weighting causes the nerve connections of white matter to appear white, and the congregations of neurons of gray matter to appear gray, while cerebrospinal fluid appears dark.

The contrast of "white matter," "gray matter'" and "cerebrospinal fluid" is reversed using T2 or T2* imaging, whereas proton-weighted imaging provides little contrast in normal subjects. Additionally, functional information (CBF, CBV, blood oxygenation) can be encoded within T1, T2, or T2.

Diffusion weighted imaging (DWI) uses very fast scans with an additional series of gradients (diffusion gradients) rapidly turned on and off. Protons from water diffusing randomly within the brain, via Brownian motion, lose phase coherence and, thus signal during application of diffusion gradients. In a brain with an acute infarction water diffusion is impaired, and signal loss on DWI sequences is less than in normal brain. DWI is the most sensitive method of detecting cerebral infarction (stroke) and works within 30 minutes of the ictus.

Contrast Enhancement

Both T1-weighted and T2-weighted images are acquired for most medical examinations; However they do not always adequately show the anatomy or pathology. The first option is to use a more sophisticated image acquisition technique such as fat suppression or chemical-shift imaging. The other is to administer a contrast agent to delineate areas of interest.

A contrast agent may be as simple as water, taken orally, for imaging the stomach and small bowel although substances with specific magnetic properties may be used. Most commonly, a paramagnetic contrast agent (usually a gadolinium compound) is given. Gadolinium-enhanced tissues and fluids appear extremely bright on T1-weighted images. This provides high sensitivity for detection of vascular tissues (e.g. tumors) and permits assessment of brain perfusion (e.g. in stroke).

There have been concerns raised recently regarding the toxicity of gadolinium-based contrast agents and their impact on persons with impaired kidney function. Special actions may be taken, such as hemodialysis following a contrast MRI scan for renally-impaired patients.

More recently, superparamagnetic contrast agents (e.g. iron oxide nanoparticles) have become available. These agents appear very dark on T2-weighted images and may be used for liver imaging - normal liver tissue retains the agent, but abnormal areas (e.g. scars, tumors) do not. They can also be taken orally, to improve visualization of the gastrointestinal tract, and to prevent water in the gastrointestinal tract from obscuring other organs (e.g. pancreas).

Diamagnetic agents such as barium sulfate have been studied for potential use in the gastrointestinal tract, but are less frequently used.

MRI VS CT

A computed tomography (CT) scanner uses X-rays, a type of ionizing radiation, to acquire its images, making it a good tool for examining tissue composed of elements of a relatively higher atomic number than the tissue surrounding them, such as bone and calcifications (calcium based) within the body (carbon based flesh), or of structures (vessels, bowel). MRI, on the other hand, uses non-ionizing radio frequency (RF) signals to acquire its images and is best suited for non-calcified tissue.

CT may be enhanced by use of contrast agents containing elements of a higher atomic number than the surrounding flesh such as iodine or barium. Contrast agents for MRI are those which have paramagnetic properties. One example is gadolinium.

Both CT and MRI scanners can generate multiple two-dimensional cross-sections (slices) of tissue and three-dimensional reconstructions. Unlike CT, which uses only X-ray attenuation to generate image contrast, MRI has a long list of properties that may be used to generate image contrast. By variation of scanning parameters, tissue contrast can be altered and enhanced in various ways to detect different features.

MRI can generate cross-sectional images in any plane (including oblique planes). CT was limited to acquiring images in the axial (or near axial) plane in the past. The scans used to be called Computed *Axial* Tomography scans (CAT scans). However, the development of multi-detector CT scanners with near-isotropic resolution, allows the CT scanner to produce data that can be retrospectively reconstructed in any plane with minimal loss of image quality.

For purposes of tumor detection and identification, MRI is generally superior. However, CT usually is more widely available, faster, much less expensive, and may be less likely to require the person to be sedated or anesthetized.

MRI is also best suited for cases when a patient is to undergo the exam several times successively in the short term, because, unlike CT, it does not expose the patient to the hazards of ionizing radiation.

Economics of MRI

MRI equipment is expensive. New 1.5 Tesla scanners often cost between $1,000,000 USD and $1,500,000 USD. New 3.0 Tesla scanners often cost between $2,000,000 and $2,300,000 USD. Construction of MRI suites can cost $500,000 USD.

For over a dozen years, MRI scanners have been significant sources of revenue for healthcare providers in the US. This is because of favorable reimbursement rates from insurers, both private and federal government programmes. Insurance reimbursement has historically been provided in two components, technical for the actual performance of the MRI scan and professional for the radiologist's review of the images and/or data.

In the US, the 2007 Deficit Reduction Act (DRA) significantly reduced reimbursement rates paid by federal insurance programmes for the technical component of many scans, shifting the economic landscape. Many private insurers have followed suit. Currently, in the US, there is increasing interest in reducing the costs associated with MRI services and simultaneously improving the ability to effectively and efficiently provide MRI examination services to larger numbers of patients with the same equipment.

The K-space Formalism

In 1983 Ljunggren and Tweig independently introduced the k-space formalism, a technique that proved invaluable in unifying different MR imaging techniques. They showed that the demodulated MR signal $S(t)$ generated by freely precessing nuclear spins in the presence of a linear magnetic field gradient G equals the Fourier transform of the effective spin density ρ_{eff} i.e.

$$S(t)=\tilde{\rho}_{\text{effective}}\left(\vec{k}(t)\right)\equiv\int d^3x\rho(\vec{x}).e^{2\pi i\vec{k}(t).\vec{x}}$$

where:

$$\vec{k}(t)=\int_0^t\vec{G}(t')dt'$$

In other words, as time progresses the signal traces out a trajectory in k-space with the velocity vector of the trajectory proportional to the vector of the applied magnetic field gradient. By the term *effective spin density* we mean the true spin density $\rho(\vec{x})$ corrected for the effects of T_1 preparation, T_2 decay, dephasing due to field inhomogeneity, flow, diffusion, etc. and any other phenomena that affect that amount of transverse magnetization available to induce signal in the RF probe.

From the basic k-space formula, it follows immediately that we reconstruct an image $I(\vec{x})$ simply by taking the inverse Fourier transform of the sampled data viz.

$$I(\vec{x})=\int d^3kS\left(\vec{k}(t)\right)\cdot e^{-2\pi i\vec{k}(t).\vec{x}}$$

Using the k-space formalism, a number of seemingly complex ideas become simple. It becomes very easy to understand the role of phase encoding (the so-called spin-warp method). In a standard spin echo or gradient echo scan, where the readout (or view) gradient is constant (e.g. G_x), a single line of k-space is scanned per RF excitation. When the phase encoding gradient is zero, the line scanned is the k_x axis. When a non-zero phase-encoding pulse is added in between the RF excitation and the commencement of the readout gradient, this line moves up or down in k-space i.e. we scan the line k_y=constant.

The k-space formalism also makes it very easy to compare different scanning techniques. In single-shot EPI, all of k-space is scanned in a single shot, following either a sinusoidal or zig-zag trajectory. Since alternating lines of k-space are scanned in opposite directions, this must be taken into account in the reconstruction.

Multi-shot EPI and fast spin echo techniques acquire only part of k-space per excitation. In each shot, a different interleaved segment is acquired, and the shots are repeated until k-space is sufficiently well-covered. Since the data at the centre of k-space represent lower spatial frequencies than the data at the edges of k-space, the T_E value for the centre of k-space determines the image's T_2 contrast.

The importance of the centre of k-space in determining image contrast can be exploited in more advanced imaging techniques. One such technique is spiral acquisition - a rotating magnetic field gradient is applied, causing the trajectory in k-space to trace out spiral out from the centre to the edge. Due to T_2 and T_2 decay the signal is greatest at the start of the acquisition, hence acquiring the centre of k-space first improves contrast to noise ratio (CNR) when compared to conventional zig-zag acquisitions, especially in the presence of rapid movement.

Since $\vec{x}$ and $\vec{k}$ are conjugate variables (with respect to the Fourier transform) we can use the Nyquist theorem to show that the step in k-space determines the field of view of the image (maximum frequency that is correctly sampled) and the maximum value of k sampled determines the resolution i.e.

$$FOV \alpha \frac{1}{\Delta k} \quad \text{Resolution } \alpha |k_{\max}|$$

(these relationships apply to each axis [*X*, *Y*, and *Z*] independently).

APPLICATION

In clinical practice, MRI is used to distinguish pathologic tissue (such as a brain tumor) from normal tissue. One advantage of an MRI scan is that it is thought to be harmless to the patient. It uses strong magnetic fields and non-ionizing radiation in the radio frequency range. Compare this to CT scans and traditional X-rays which involve doses of ionizing radiation and may increase the risk of malignancy, especially in a fetus.

While CT provides good spatial resolution (the ability to distinguish two structures an arbitrarily small distance from each other as separate), MRI provides comparable resolution with far better contrast resolution (the ability to distinguish the differences between two arbitrarily similar but not identical tissues). The basis of this ability is the complex library of *pulse sequences* that the modern medical MRI scanner includes, each of which is optimized to provide *image contrast* based on the chemical sensitivity of MRI.

With particular values of the *echo time* (TE) and the *repetition time* (TR), which are basic parameters of image acquisition, a sequence will take on the property of T2-weighting. On a T2-weighted scan, fat-, water- and fluid-containing tissues are bright (most modern T2 sequences are actually *fast T2* sequences). Damaged tissue tends to develop edema, which makes a T2-weighted sequence sensitive for pathology, and generally able to distinguish pathologic tissue from normal tissue. With the addition of an additional radio frequency pulse and additional manipulation of the magnetic gradients, a T2-weighted sequence can be converted to a FLAIR sequence, in which free water is now dark, but edematous tissues remain bright. This sequence in particular is currently the most sensitive way to evaluate the brain for demyelinating diseases, such as multiple sclerosis.

The typical MRI examination consists of 5-20 sequences, each of which are chosen to provide a particular type of information about the subject tissues. This information is then synthesized by the interpreting physician.

Specialized MRI Scans

Diffusion MRI

Diffusion MRI measures the diffusion of water molecules in biological tissues. In an isotropic medium water molecules naturally move randomly according to Brownian motion. In biological tissues however, the diffusion may be anisotropic. A molecule inside the axon of a neuron has a low probability of crossing the myelin membrane. Therefore the molecule will move principally along the axis of the neural fibre. If we know that molecules in a particular voxel diffuse principally in one direction we can make the assumption that the majority of the fibres in this area are going parallel to that direction.

The recent development of diffusion tensor imaging (DTI) enables diffusion to be measured in multiple directions and the fractional anisotropy in each direction to be calculated for each voxel. This enables researchers to make brain maps of fibre directions to examine the connectivity of different regions in the brain (using tractography) or to examine areas of neural degeneration and demyelination in diseases like Multiple Sclerosis.

Another application of diffusion MRI is diffusion-weighted imaging (DWI). Following an ischemic stroke, DWI is highly sensitive to the changes occurring in the lesion. It is speculated that increases in restriction (barriers) to water diffusion, as a result of cytotoxic edema (cellular swelling), is responsible for the increase in signal on a DWI scan. Other theories, including acute changes in cellular permeability and loss of energy-dependent (ATP) cytoplasmic streaming, have been proposed to explain the phenomena.

The DWI enhancement appears within 5-10 minutes of the onset of stroke symptoms (as compared with computed tomography, which often does not detect changes of acute infarct for up to 4-6 hours) and remains for up to two weeks. CT, due to its insensitivity to acute ischemia, is typically employed to rule out hemorrhagic stroke, which would entirely prevent the use of tissue plasminogen activator (tPA). Further, coupled with scans sensitized to cerebral perfusion, researchers can highlight regions of "perfusion/diffusion mismatch" that may indicate regions capable of salvage by reperfusion therapy.

Finally, it has been proposed that diffusion MRI may be able to detect minute changes in extracellular water diffusion and therefore could be used as a tool for fMRI. The nerve cell body enlarges when it conducts an action potential, hence restricting extracellular water molecules from diffusing naturally. Although this process works in theory, evidence is only moderately convincing.

Like many other specialized applications, this technique is usually coupled with a fast image acquisition sequence, such as echo planar imaging sequence.

Magnetic Resonance Angiography

Magnetic resonance angiography (MRA) is used to generate pictures of the arteries in order to evaluate them for stenosis or aneurysms (vessel wall dilatations, at risk of rupture).

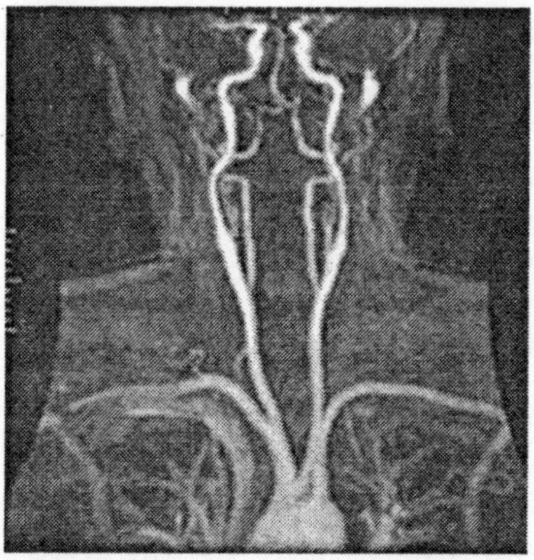

Fig. Magnetic Resonance Angiography

MRA is often used to evaluate the arteries of the neck and brain, the thoracic and abdominal aorta, the renal arteries, and the legs (called a "run-off").

A variety of techniques can be used to generate the pictures, such as administration of a paramagnetic contrast agent (gadolinium) or using a technique known as "flow-related enhancement" (e.g. 2D and 3D time-of-flight sequences), where most of the signal on an image is due to blood which has recently moved into that plane. Magnetic resonance venography (MRV) is a similar procedure that is used to image veins.

In this method the tissue is now excited inferiorly while signal is gathered in the plane immediately superior to the excitation plane, and thus imaging the venous blood which has recently moved from the excited plane.

Magnetic Resonance Spectroscopy

In vivo ('in the living organism') magnetic resonance spectroscopy (MRS), also known as MRSI (MRS imaging) and volume selective NMR spectroscopy, is a technique which combines the spatially-addressable nature of MRI with the spectroscopically-rich information obtainable from NMR.

That is to say, MRI allows one to study a particular region within an organism or sample, but gives relatively little information about the chemical or physical nature of that region (its chief value is in being able to distinguish the properties of that region, how much fat or water is present, relative to those of surrounding regions). MR spectroscopy, however, provides a wealth of information about other biological chemicals ('metabolites') within that region, as would an NMR spectrum of that region.

Functional MRI

A fMRI scan showing regions of activation in orange, including the primary visual cortex.

Functional MRI (fMRI) measures signal changes in the brain that are due to changing neural activity. The brain is scanned at low resolution but at a rapid rate (typically once every 2-3 seconds). Increases in neural activity cause changes in the MR signal via T2* changes; this mechanism is referred to as the BOLD (blood-oxygen-level dependent) effect.

Increased neural activity causes an increased demand for oxygen, and the vascular system actually overcompensates for this, increasing the amount of oxygenated hemoglobin relative to deoxygenated hemoglobin.

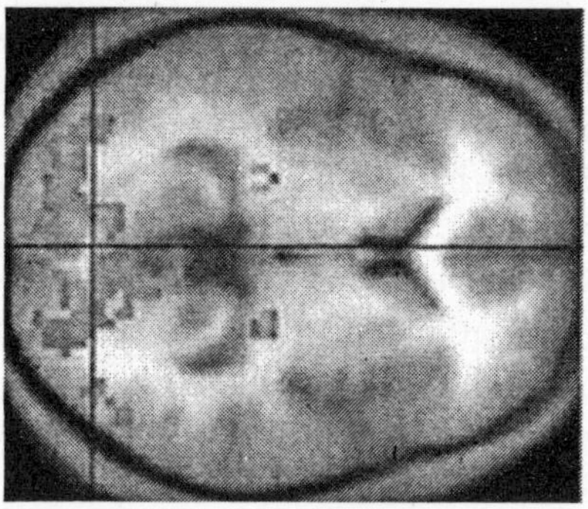

Because deoxygenated hemoglobin attenuates the MR signal, the vascular response leads to a signal increase that is related to the neural activity. The precise nature of the relationship between neural activity and the BOLD signal is a subject of current research. The BOLD effect also allows for the generation of high resolution 3D maps of the venous vasculature within neural tissue.

While BOLD signal is the most common method employed for neuroscience studies in human subjects, the flexible nature of MR imaging provides means to sensitize the signal to other aspects of the blood supply. Alternative techniques employ arterial spin labeling (ASL) or weight the MRI signal by cerebral blood flow (CBF) and cerebral blood volume (CBV). The CBV method requires injection of a class of MRI contrast agents that are now in human clinical trials. Because this method has been shown to be far more sensitive than the BOLD technique in preclinical studies, it may potentially expand the role of fMRI in clinical applications. The CBF method provides more quantitative information than the BOLD signal, albeit at a significant loss of detection sensitivity.

Interventional MRI

The lack of harmful effects on the patient and the operator make MRI well-suited for "interventional radiology", where the images produced by a MRI scanner are used to guide minimally-invasive procedures.

Radiation Therapy Simulation

Because of MRI's superior imaging of soft tissues, it is now being utilized to specifically locate tumors within the body in preparation for radiation therapy treatments. For therapy simulation, a patient is placed in specific, reproducible, body position and scanned. The MRI system then computes the precise location, shape and orientation of the tumor mass, correcting for any spatial distortion inherent in the system. The patient is then marked or tattooed with points which, when combined with the specific body position, will permit precise triangulation for radiation therapy.

Current Density Imaging

Current density imaging (CDI) endeavors to use the phase information from images to reconstruct current densities within a subject. Current density imaging works because electrical currents generate magnetic fields, which in turn affect the phase of the magnetic dipoles during an imaging sequence. To date no successful CDI has been performed using biological currents, but several studies have been published which involve applied currents through a pair of electrodes.

Magnetic Resonance Guided Focused Ultrasound

In MRgFUS therapy, ultrasound beams are focused on a tissue - guided and controlled using MR thermal imaging - and due to the significant energy deposition at the focus, temperature within the tissue rises to more than 65°C, completely destroying it.

This technology can achieve precise "ablation" of diseased tissue. MR imaging provides a three-dimensional view of the target tissue, allowing for precise focusing of ultrasound energy.

The MR imaging provides quantitative, real-time, thermal images of the treated area. This allows the physician to ensure that the temperature generated during each cycle of ultrasound energy is sufficient to cause thermal ablation within the desired tissue and if not, to adapt the parameters to ensure effective treatment.

Multinuclear Imaging

Hydrogen is the most frequently imaged nucleus in MRI because it is present in biological tissues in great abundance. However, any nucleus which has a net nuclear spin could potentially be imaged with MRI. Such nuclei include helium-3, carbon-13, fluorine-19, oxygen-17, sodium-23, phosphorus-31 and xenon-129. ^{23}Na and ^{31}P are naturally abundant in the body, so can be imaged directly.

Gaseous isotopes such as ^{3}He or ^{129}Xe must be hyperpolarized and then inhaled as their nuclear density is too low to yield a useful signal under normal conditions. ^{17}O, ^{13}C and ^{19}F can be administered in sufficient quantities in liquid form (e.g. ^{17}O-water, ^{13}C-glucose solutions or perfluorocarbons) that hyperpolarization is not a necessity. Multinuclear imaging is primarily a research technique at present.

However, potential applications include functional imaging and imaging of organs poorly seen on ^{1}H MRI (e.g. lungs and bones) or as alternative contrast agents.

Inhaled hyperpolarized ^{3}He can be used to image the distribution of air spaces within the lungs. Injectable solutions containing ^{13}C or stabilized bubbles of hyperpolarized ^{129}Xe have been studied as contrast agents for angiography and perfusion imaging. ^{31}P can potentially provide information on bone density and structure, as well as functional imaging of the brain.

Experimental MRI Techniques

Currently there is active research in several new MRI technologies like magnetization transfer MRI (MT-MRI), diffusion tensor MRI (DT-MRI), and proton MR spectroscopy, plus recent research in to Dendrimer-enhanced MRI as a diagnostic and prognostic biomarker of sepsis-induced acute renal failure.

SAFETY

Implants and foreign bodies: Pacemakers are generally considered an absolute contraindication towards MRI scanning, though highly specialized protocols have been developed to permit scanning of select pacing devices. Several cases of arrhythmia or death have been reported in patients with pacemakers who have undergone MRI scanning without appropriate precautions. Other electronic implants have varying contraindications, depending upon scanner technology, implant properties, scanning protocols and anatomy being imaged.

Though pacemakers receive significant attention, it should also be noted that many other forms of medical or biostimulation implants may be contraindicated for MRI scans. These may include Vagus nerve stimulators, implantable cardioverter-defibrillators (ICD), loop recorders, insulin pumps, cochlear implants, deep brain stimulators and many others.

Medical device patients should always present complete information (manufacturer, model, serial number and date of implantation) about all implants to both the referring physician and to the radiologist or technologist before entering the room for the MRI scan.

While these implants pose a current problem, scientist and manufacturers are working on improved designs which will further minimize the risks that MRI scans pose to medical device operations. One such development in the works is a nano coating for implants intended to screen them from the radio frequency waves, helping to make MRI exams available to patients currently prohibited from receiving them.

Ferromagnetic foreign bodies (e.g. shell fragments), or metallic implants (e.g. surgical prostheses, aneurysm clips) are also potential risks, and safety aspects need to be considered on an individual basis. Interaction of the magnetic and radio frequency fields with such objects can lead to: trauma due to movement of the object in the magnetic field, thermal injury from radio-frequency induction heating of the object, or failure of an implanted device. These issues are especially problematic when dealing with the eye. Most MRI centres require an orbital x-ray be performed on anyone who suspects they may have small metal fragments in their eyes, perhaps from a previous accident, something not uncommon in metalworking.

Because of its non-ferromagnetic nature and poor electrical conductivity, titanium and its alloys are useful for long term implants and surgical instruments intended for use in image-guided surgery.

In particular, not only is titanium safe from movement from the magnetic field, but artifacts around the implant are less frequent and less severe than with more ferromagnetic materials e.g. stainless steel. Artifacts from metal frequently appear as regions of empty space around the implant - frequently called 'black-hole artifact' e.g. a 3mm titanium alloy coronary stent may appear as a 5mm diameter region of empty space on MRI, whereas around a stainless steel stent, the artifact may extend for 10-20 mm or more.

In 2006, a new classification system for implants and ancillary clinical devices has been developed by ASTM International and is now the standard supported by the US Food and Drug Administration:

MR-Safe: The device or implant is completely non-magnetic, non-electrically conductive, and non-RF reactive, eliminating all of the primary potential threats during an MRI procedure.

MR-Conditional: A device or implant that may contain magnetic, electrically conductive or RF-reactive components that is safe for operations in proximity to the MRI, provided the conditions for safe operation are defined and observed.

MR-Unsafe: Nearly self-explanatory, this category is reserved for objects that are significantly ferromagnetic and pose a clear and direct threat to persons and equipment within the magnet room.

In the case of pacemakers, the risk is thought to be primarily RF induction in the pacing electrodes/wires causing inappropriate pacing of the heart, rather than the magnetic field affecting the pacemaker itself. Much research and development is being undertaken, and many tools are being developed in order to predict the effects of the RF fields inside the body.

Projectile or Missile Effect

As a result of the very high strength of the magnetic field needed to produce scans (frequently up to 60,000 times the earth's own magnetic field effects), there are several incidental safety issues addressed in MRI facilities.

Missile-effect accidents, where ferromagnetic objects are attracted to the centre of the magnet, have resulted in injury and death. A video simulation of a fatal projectile effect accident illustrates the extreme power that contemporary MRI equipment can exert on ferromagnetic objects.

In order to help reduce the risks of projectile accidents, ferrous objects and devices are typically prohibited in proximity to the MRI scanner, with non ferro-magnetic versions of many tools and devices typically retained by the scanning facility.

Patients undergoing MRI examinations are required to remove all metallic objects, often by changing into a gown or 'scrubs'.

New ferromagnetic-only detection devices (ferromagnetic detectors) are proving highly effective in supplementing conventional screening techniques in many leading hospitals and imaging centres and are now recommended by the American College of Radiology's *Guidance Document for Safe MR Practices: 2007*.

The magnetic field and the associated risk of missile-effect accidents remains a permanent hazard — as superconductive MRI magnets retain their magnetic field, even in the event of a power outage.

Radio Frequency Energy

A powerful radio transmitter is needed for excitation of proton spins. This can heat the body significantly, with the risk of hyperthermia in patients, particularly the obese or patients with thermoregulation disorders. Several countries have issued restrictions on the maximum specific absorption rate that a scanner may produce.

Peripheral Nerve Stimulation

The rapid switching (on and off) of the magnetic field gradients needed for imaging is capable of causing nerve stimulation. Volunteers report a twitching sensation when exposed to rapidly switched fields, particularly in their extremities.

The reason the peripheral nerves are stimulated is that the changing field increases with distance from the centre of the gradient coils (which more or less coincides with the centre of the magnet).

Note however that when imaging the head, the heart is far off-centre and induction of even a tiny current into the heart must be avoided at all costs. Although PNR was not a problem for the slow, weak gradients used in the early days of MRI, the strong, rapidly-switched gradients used in techniques such as EPI, fMRI, diffusion MRI, etc. are indeed capable of inducing PNR.

Acoustic Noise

Loud noises and vibrations are produced by forces resulting from rapidly switched magnetic gradients interacting with the main magnetic field, in turn causing minute expansions and contractions of the coil itself. This is most marked with high-field machines and rapid-imaging techniques in which sound intensity can reach 130 dB (equivalent to a jet engine at take-off).

Appropriate use of ear protection is essential for anyone inside the MRI scanner room during the examination.

Cryogens

As described in 'Scanner Construction And Operation', many MRI scanners rely on cryogenic liquids to enable superconducting capabilities of the electromagnetic coils within. Though the cryogenic liquids most frequently used are non-toxic, their physical properties present specific hazards.

An emergency shut-down of a superconducting electromagnet, an operation known as "quenching", involves the rapid boiling of liquid helium from the device.

If the rapidly expanding helium cannot be dissipated though an external vent, sometimes referred to as 'quench pipe', it may be released into the scanner room where it may cause displacement of the oxygen and present a risk of asphyxiation.

Liquid helium, the most commonly used cryogen in MRI, undergoes near explosive expansion as it changes from liquid to a gaseous state. Rooms built in support of superconducting MRI equipment should be equipped with pressure relief mechanisms and an exhaust fan, in addition to the required quench pipe.

Since a quench results in rapid loss of all cryogens in the magnet, recommissioning the magnet is extremely expensive and time-consuming. Spontaneous quenches are uncommon, but can occur at any time. Quenches may also be triggered by equipment malfunction, improper cryogen fill technique, contaminates inside the cryostat, or extreme magnetic or vibrational disturbances.

Contrast Agents

The most frequently used intravenous contrast agents are based on chelates of gadolinium. In general, these agents have proved safer than the iodinated contrast agents used in X-ray radiography or CT. Anaphylactoid reactions are rare occurring in approx 0.03-0.1%. Of particular interest is the lower incidence of nephrotoxicity, compared with iodinated agents, when given at usual doses—this has made contrast-enhanced MRI scanning an option for patients with renal impairment, who would otherwise not be able to undergo contrast-enhanced CT.

Although gadolinium agents have proved useful for patients with renal impairment, in patients with severe renal failure requiring dialysis there is a risk of a rare but serious illness, nephrogenic systemic fibrosis, that may be linked to the use of certain gadolinium-containing agents: the most frequently linked is gadodiamide, but other agents have been linked too. Although a causal link has not been definitively established, current guidelines are that dialysis patients should only receive gadolinium agents where essential, and that dialysis should be performed as soon as possible after the scan is complete, in order to remove the agent from the body promptly. In Europe where more gadolinium-containing agents are available, a classification of agents according to potential risks has been released.

Pregnancy

No harmful effects of MRI on the fetus have been demonstrated. In particular, MRI avoids the use of ionizing radiation, to which the fetus is particularly sensitive. However, as a precaution, current guidelines recommend that pregnant women undergo MRI only when essential. This is particularly the case during the first trimester of pregnancy, as organogenesis takes place during this period.

The concerns in pregnancy are the same as for MRI in general, but the fetus may be more sensitive to the effects—particularly to heating and to noise. However, one additional concern is the use of contrast agents; gadolinium compounds are known to cross the placenta and enter the fetal bloodstream, and it is recommended that their use be avoided.

Despite these concerns, MRI is rapidly growing in importance as a way of diagnosing and monitoring congenital defects of the fetus because it can provide more diagnostic information than ultrasound and it lacks the ionizing radiation of CT. MRI without contrast is the imaging mode of choice for pre-surgical, in-utero diagnosis and evaluation of fetal tumors, primarily teratomas, facilitating open fetal surgery, other fetal interventions, and planning for procedures (such as the EXIT procedure) to safely deliver and treat babies whose defects would otherwise be fatal.

Claustrophobia and Discomfort

Due to the construction of MRI scanners, they are potentially unpleasant in which to lie. The part of the body being imaged needs to lie at the centre of the magnet (which is often a long, narrow tube). Because scan times may be long (perhaps one hour), people with even mild claustrophobia are often unable to tolerate an MRI scan without management.

Management may include:

- Advance preparation
  - Visiting the scanner to see the room and practice lying on the table
  - Visualization techniques
  - Chemical sedation
  - General anesthesia
- Coping while inside the scanner
  - Holding a "panic button"
  - Listening to music on headphones or watching a movie with a Head-mounted display while in the machine
- Modified scanner designs
  - Upright MRIs (made exclusively by FONAR)
  - Open-bore design scanners.

Though open MRIs have increased in popularity as of late, they produce inferior scan quality because they operate at lower magnetic fields than closed MRIs.

For babies and children, chemical sedation or general anesthesia are the norm. These MRI subjects are too young to be instructed to hold still during the scanning session. Obese patients and pregnant women may find the MRI machine to be a tight fit, and even some claustrophobics may find the experience intolerable without sedation. Pregnant women may also have difficulty lying on their backs for an hour or more without moving.

The noise associated with the operation of an MRI scanner (especially the audible noise associated with the gradient pulses applied to the subject) can also exacerbate the discomfort associated with the procedure.

The European Physical Agents Directive

The European Physical Agents (Electromagnetic Fields) Directive is European legislation that has been adopted in

European legislature. By 2008 each individual state within the European Union must include this directive in its own law.

The directive applies to occupational exposure to electromagnetic fields (not medical exposure) and was intended to limit workers' acute exposure to strong electromagnetic fields, as may be found near electricity substations, radio or television transmitters or industrial equipment.

However, the regulations impact significantly on MRI, with separate sections of the regulations limiting exposure to static magnetic fields, changing magnetic fields and radio frequency energy. Field strength limits are given which may not be exceeded for any period of time. An employer may commit a criminal offense by allowing a worker to exceed an exposure limit if that is how the Directive is implemented in a particular Member State.

The Directive is based on the international consensus of established effects of exposure to electromagnetic fields, and in particular the advice of the European Commissions's advisor, the International Commission on Non-Ionizing Radiation Protection (ICNIRP). The aims of the Directive, and the ICNIRP guidelines upon which it is based, are to prevent exposure to potentially harmful fields. The actual limits in the Directive are very similar to the limits advised by the Institute of Electrical and Electronics Engineers, with the exception of the frequencies produced by the gradient coils, where the IEEE limits are significantly higher.

Many Member States of the EU already have either specific EMF regulations or (as in the UK) a general requirement under workplace health and safety legislation to protect workers against electromagnetic fields. In almost all cases the existing regulations are aligned with the ICNIRP limits so that the Directive should, in theory, have little impact on any employer already meeting their legal responsibilities.

The introduction of the Directive has brought to light an existing potential issue with occupational exposures to MRI fields. There are at present very few data on the number or types of MRI practice that might lead to exposures in excess of the levels of the Directive. There is a justifiable concern amongst MRI practitioners that if the Directive were to be enforced more vigourously than existing legislation, the use of MRI might be restricted, or working practices of MRI personnel might have to change.

In the initial draft a limit of static field strength to 2 T was given. This has since been removed from the regulations, and whilst it is unlikely to be restored as it was without a strong justification, some restriction on static fields may be reintroduced after the matter has been considered more fully by ICNIRP.

The effect of such a limit might be to restrict the installation, operation and maintenance of MRI scanners with magnets of 2 T and stronger. As the increase in field strength has been instrumental in developing higher resolution and higher performance scanners, this would be a significant step back. This is why it is unlikely to happen without strong justification.

Individual government agencies and the European Commission have now formed a working group to examine the implications on MRI and to try to address the issue of occupational exposures to electromagnetic fields from MRI.

Chapter 6

Electroencephalography

SOURCE OF EEG ACTIVITY

Electroencephalography (EEG) is the measurement of electrical activity produced by the brain as recorded from electrodes placed on the scalp. Just as the activity in a computer can be perceived on multiple different levels, from the activity of individual transistors to the function of applications, so can the electrical activity of the brain be described on relatively small to relatively large scales. At one end are action potentials in a single axon or currents within a single dendrite, and at the other end is the activity measured by the scalp EEG. The data measured by the scalp EEG are used for clinical and research purposes.

A technique similar to the EEG is intracranial EEG (icEEG), also referred to as subdural EEG (sdEEG) and electrocorticography (ECoG). These terms refer to the recording of activity from the surface of the brain (rather than the scalp). Because of the filtering characteristics of the skull and scalp, icEEG activity has a much higher spatial resolution than surface EEG. Scalp EEG measures summated activity of post-synaptic currents. An action potential in a pre-synaptic axon causes the release of neurotransmitter into the synapse. The neurotransmitter diffuses across the synaptic cleft and binds to receptors in a post-synaptic dendrite.

The activity of many types of receptors results in a flow of ions into or out of the dendrite. This results in compensatory currents in the extracellular space. It is these extracellular currents which are responsible for the generation of EEG voltages. The EEG is not sensitive to axonal action potentials. While it is post-synaptic potentials which generate the EEG signal, it is not possible to determine the activity within a single dendrite or neuron from the scalp EEG. Rather, surface EEG is the summation of the synchronous activity of thousands of neurons that have similar spatial orientation, radial to the scalp.

Currents that are tangential to the scalp are not picked up by the EEG. The EEG therefore benefits from the parallel, radial arrangement of apical dendrites in the cortex. Because voltage fields fall off with the fourth power of the radius, activity from deep sources is more difficult to detect than currents near the skull. Scalp EEG activity is comprised of multiple oscillations. These have a different characteristic frequencies, spatial distributions and associations with different states of brain functioning (such as awake vs. asleep).

These oscillations represent synchronized activity over a network of neurons. The neuronal network underlying some of these oscillations are understood (such as the thalomocortical resonance underlying sleep spindles), while many others are not (e.g., the system that generates the posterior basic rhythm still defies understanding).

Clinical Use

A routine clinical EEG recording typically lasts 20-40 minutes. During this time, it is common to perform different "activation procedures" which may evoke different activity than is seen during the resting awake state.

These activation procedures include sleep, intermittent photic stimulation with a strobe light, hyperventilation and eye closure. When a routine EEG is done in a patient with suspected or known epilepsy, often it is to look for inter-ictal discharges (i.e., abnormal activity resulting from "brain irritability" that shows a possible predisposition to epileptic seizures.

In certain cases, video-EEG monitoring may be required. This is simultaneous recording of EEG and time-locked video/ audio. Video-EEG monitoring may involve inpatient admissions for days to weeks. During these admissions, a patient's anti-epilpetic medications are often titrated off so that seizures can be recorded. The EEG appearance of the onset of the seizure can provide significantly more definitive information about the patient's epilepsy than inter-ictal recordings can in many cases.

Continuous EEG monitoring typically involves the use of a portable EEG machine connected to an ICU patient to look for seizure activity that is not apparent clinically (i.e., in the patient's mental status or by observing his/her movements). When patients are put into medically-induced comas, the EEG pattern may be used as measure of depth of coma, and the medication may be titrated to an EEG end-point. "Amplitude-integrated EEG" refers to a specific representation of the EEG signal that is used with continuous monitoring of cerebral function in some neonatal ICUs.

These various forms of EEG recording can be used in the following clinical situations:

- To distinguish epileptic seizures from other types of spells, such as psychogenic non-epileptic seizures, syncope (fainting), sub-cortical movement disorders and migraine variants.
- To characterize seizures for the purposes of treatment
- To localize the region of brain from which a seizure originates for work-up of possible seizure surgery
- To monitor for non-convulsive seizures/non-convulsive status epilepticus
- To differentiate "organic" encephalopathy or delirium from primary psychiatric syndromes such as catatonia
- Monitoring depth of anesthesia
- As an indirect indicator of cerebral perfusion in carotid endarterectomy
- To serve as an adjunct test of brain death
- To prognosticate, in certain instances, in patients with coma

Use of the quantitative EEG (mathematical measurement of aspects of the EEG signal) in primary psychiatric, behavioural and learning disorders is somewhat controversial.

Research use

There are a number of benefits to using EEG in neuroscience research. One is that EEG is non-invasive to the research subject.

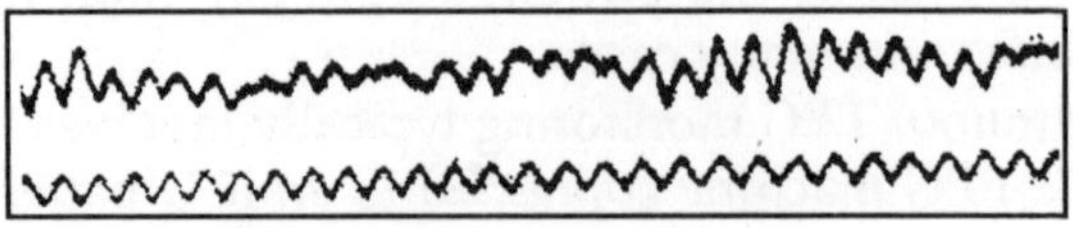

Fig. An Early EEG Recording

Furthermore, the need for the subject to hold still is perhaps less stringent than in functional magnetic resonance imaging (fMRI). Another benefit is that many applications of EEG record spontaneous brain activity, and the subject does not need to be able to cooperate with the research (e.g., as is necessary in the behavioural testing of neuropsychology).

Also, EEG has a high temporal resolution compared to techniques such as fMRI and is capable of detecting changes in electrical activity in the brain on a millisecond time scale. Much of the cognitive research conducted with EEG uses the event-related potential (ERP) technique. Most ERP paradigms involve a subject being provided a stimulus to react to either overtly or covertly.

There are often at least two conditions that vary in some manner of interest to the researcher.

As this stimulus-response is going on, an EEG is being recorded from the subject. The ERP is obtained by averaging the EEG signal from each of the trials within a certain condition; averages from one stimulus-response condition can then be compared with averages from the other stimulus-response condition(s).

Method

In conventional scalp EEG, the recording is obtained by placing electrodes on the scalp with a conductive gel or paste, usually after preparing the scalp area by light abrasion to reduce impedance due to dead skin cells. The technique has been advanced by the use of carbon nanotubes to penetrate the outer layers of the skin for improved electrical contact.

The sensor is known as ENOBIO; however, this technique is not in common research or clinical use. Many systems typically use electrodes, each of which is attached to an individual wire. Some systems use caps or nets into which electrodes are embedded; this is particularly common when high-density arrays of electrodes are needed. Electrode locations and names are specified by the International 10–20 system for most clinical and research applications (except when high-density arrays are used).

This system ensures that the naming of electrodes is consistent across laboratories. In most clinical applications, 19 recording electrodes (plus ground and system reference) are used. A smaller number of electrodes are typically used when recording EEG from neonates. Additional electrodes can be added to the standard set-up when a clinical or research application demands increased spatial resolution for a particular area of the brain.

High-density arrays (typically via cap or net) can contain up to 256 electrodes more-or-less evenly spaced around the scalp. Each electrode is connected to one input of a differential amplifier (one amplifier per pair of electrodes); a common system reference electrode is connected to the other input of each differential amplifier. These amplifiers amplify the voltage between the active electrode and the reference (typically 1,000–100,000 times, or 60–100 dB of voltage gain). In analog EEG, the signal is then filtered, and the EEG signal is output as the deflection of pens as paper passes underneath.

Most EEG systems these days, however, are digital, and the amplified signal is digitized via an analog-to-digital converter, after being passed through an anti-aliasing filter. Analog-to-digital sampling typically occurs at 256-512 Hz in clinical scalp EEG; sampling rates of up to 10 kHz are used in some research applications. The digital EEG signal is stored electronically and can be filtered for display. Typical settings for the high-pass filter and a low-pass filter are 0.5-1 Hz and 35–70 Hz, respectively.

The high-pass filter typically filters out slow artifact, such as electrogalvanic signals and movement artifact, whereas the low-pass filter filters out high-frequency artifacts, such as electromyographic signals.

An additional notch filter is typically used to remove artifact caused by electrical power lines. As part of an evaluation for epilepsy surgery, it may be necessary to insert electrodes near the surface of the brain, under the surface of the dura mater. This is accomplished via burr hole or craniotomy.

This is referred to variously as "electrocorticography (ECoG)", "intracranial EEG (I-EEG)" or "sub-dural EEG (SD-EEG)". Depth electrodes may also be placed into brain structures, such as the amygdala or hippocampus, structures which are common epileptic foci and may not be "seen" clearly by scalp EEG. The electrocorticographic signal is processed in the same manner as digial scalp EEG, with a couple of caveats.

ECoG is typically recorded at higher sampling rates than scalp EEG because of the requirements of Nyquist theorem—the sub-dural signal is composed of a higher predominance of higher frequency components. Also, many of the artifacts which affect scalp EEG do not impact ECoG, and therefore display filtering is often not needed.

A typical adult human EEG signal is about 10-100 μV in amplitude when measured from the scalp and is about 10–20 mV when measured from subdural electrodes. Since an EEG voltage signal represents a difference between the voltages at two electrodes, the display of the EEG for the reading encephalographer may be set up in one of several ways. The representation of the EEG channels is referred to as a *montage.*

BIPOLAR MONTAGE

Each channel (i.e., waveform) represents the difference between two adjacent electrodes. The entire montage consists of a series of these channels. The channel "Fp1-F3" represents the difference in voltage between the Fp1 electrode and the F3 electrode. The next channel in the montage, "F3-C3," represents the voltage difference between F3 and C3, and so on through the entire array of electrodes.

REFERENTIAL MONTAGE

Each channel represents the difference between a certain electrode and a designated reference electrode. There is no standard position at which this reference is always placed; it is, however, at a different position than the "recording" electrodes. Midline positions are often used because they do not amplify the signal in one hemisphere vs. the other. Another popular reference is "linked ears," which is a physical or mathematical average of electrodes attached to both earlobes or mastoids.

AVERAGE REFERENCE MONTAGE

The outputs of all of the amplifiers are summed and averaged, and this averaged signal is used as the common reference for each channel.

LAPLACIAN MONTAGE

Each channel represents the difference between an electrode and a weighted average of the surrounding electrodes. When analog EEGs are used, the technologist switches between montages during the recording in order to highlight or better characterize certain features of the EEG.

With digital EEG, all signals are typically digitized and stored in a particular (usually referential) montage; since any montage can be constructed mathematically from any other, the EEG can be viewed by the electroencephalographer in any display montage that is desired.

Limitations

EEG has several limitations. Most important is its poor spatial resolution. EEG is most sensitive to a particular set of post-synaptic potentials: those which are generated in superficial layers of the cortex, on the crests of gyri directly abutting the skull and radial to the skull. Dendrites which are deeper in the cortex, inside sulci, are in midline or deep structures (such as the cingulate gyrus or hippocampus) or produce currents which are tangential to the skull have far less contribution to the EEG signal.

The meninges, cerebrospinal fluid and skull "smear" the EEG signal, obscuring its intracranial source. It is mathematically impossible to reconstruct a unique intercranial current source for a given EEG signal, as some currents produce potentials that cancel each other out. This is referred to as the inverse problem. However, much work has been done to produce remarkably good estimates of, at least, a localized electric dipole that represents the recorded currents.

Advantages

EEG has several strong sides as a tool of exploring brain activity; its time resolution is very high (on the level of a single millisecond). Other methods of looking at brain activity, such as PET and fMRI have time resolution between seconds and minutes.

EEG measures the brain's electrical activity directly, while other methods record changes in blood flow (e.g., SPECT, fMRI) or metabolic activity. (e.g., PET), which are indirect markers of brain electrical activity.

EEG can be used simultaneously with fMRI so that high-temporal-resolution data can be recorded at the same time as high-spatial-resolution data, however, since the data derived from each occurs over a different time course, the data sets do not necessarily represent the exact same brain activity.

There are technical difficulties associated with combining these two modalities, including the need to remove RF pulse artifact and ballistocardiographic artifact (a results from the movement of pulsed blood) from the EEG.

Furthermore, currents can be induced in moving EEG electrode wires due to the magnetic field of the MRI. EEG can be recorded at the same time as MEG so that data from these complimentary high-time-resolution techniques can be combined.

Normal Activity

The EEG typically described in terms of:

- Rhythmic activity
- Transients.

The rhythmic activity is divided into bands by frequency.

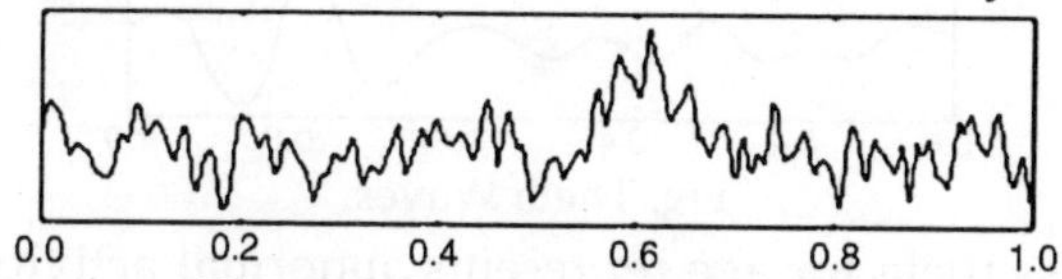

Fig. One Second of EEG Signal

To some degree, these frequency bands are a matter of nomenclature (i.e., any rhythmic activity between 8-12 Hz can be described as "alpha"), but these designations arose because rhythmic activity within a certain frequency range was noted to have a certain distribution over the scalp or a certain biological significance.

Most of the cerebral signal observed in the scalp EEG falls in the range of 1-20 Hz.

Delta Frequency

Delta is the frequency range up to 3 Hz. It tends to be the highest in amplitude and the slowest waves. It is seen normally in adults in slow wave sleep.

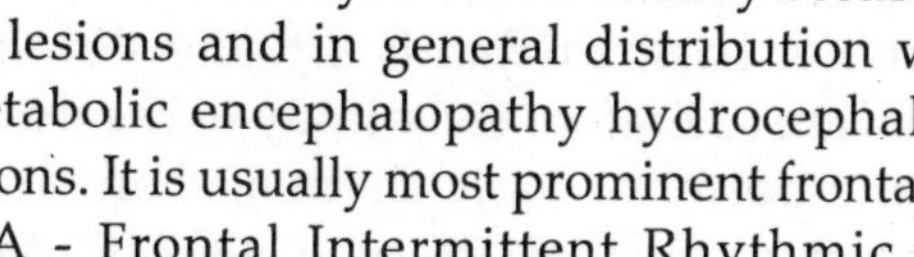

Fig. Delta Waves.

It is also seen normally in babies. It may occur focally with subcortical lesions and in general distribution with diffuse lesions, metabolic encephalopathy hydrocephalus or deep midline lesions. It is usually most prominent frontally in adults (e.g. FIRDA - Frontal Intermittent Rhythmic Delta) and posteriorly in chilldren e.g. OIRDA - Occipital Intermittent Rhythmic Delta).

Theta Frequency

Theta is the frequency range from 4 Hz to 7 Hz. Theta is seen normally in young children. It may be seen in drowsiness or arousal in older children and adults; it can also be seen in meditation.

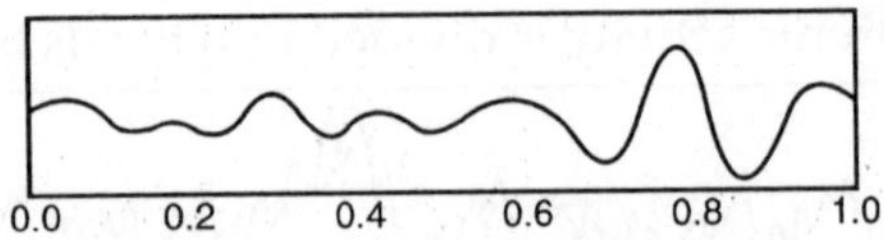

Fig. Theta Waves.

Excess theta for age represents abnormal activity. It can be seen as a focal disturbance in focal subcortical lesions; it can be seen in generalized distribution in diffuse disorder or metabolic encephalopathy or deep midline disorders or some instances of hydrocephalus.

Aa Frequency

Alpha is the frequency range from 8 Hz to 12 Hz. Hans Berger named the first rhythmic EEG activity he saw, the "alpha wave."

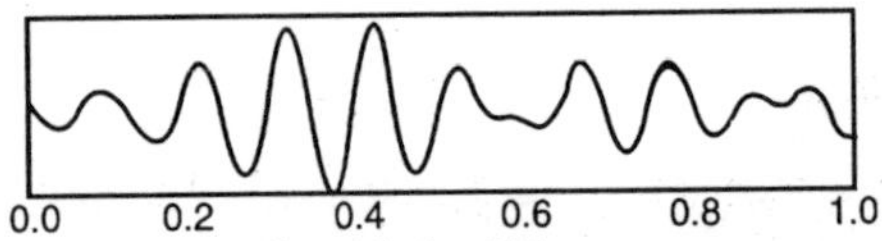

Fig. Alpha Waves.

This is activity in the 8-12 Hz range seen in the posterior regions of the head on both sides, being higher in amplitude on the dominant side. It is brought out by closing the eyes and by relaxation. It was noted to attenuate with eye opening or mental exertion. This activity is now referred to as "posterior basic rhythm," the "posterior dominant rhythm" or the "posterior alpha rhythm."

The posterior basic rhythm is actually slower than 8 Hz in young children (therefore technically in the theta range). In addition to the posterior basic rhythm, there are two other normal alpha rhythms that are typically discussed: the mu rhythm and a temporal "third rhythm". Alpha can be abnormal; an EEG that has diffuse alpha occurring in coma and is not responsive to external stimuli is referred to as "alpha coma".

Mu Frequency

Mu rhythm is alpha-range activity that is seen over the sensorimotor cortex.

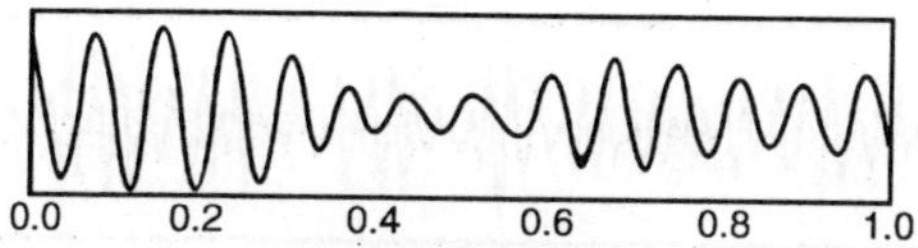

Fig. Mu Rhythm.

It characteristically attenuates with movement of the contralateral arm (or mental imagery of movement of the contralateral arm).

Beta Frequency

Beta is the frequency range from 12 Hz to about 30 Hz. It is seen usually on both sides in symmetrical distribution and is most evident frontally.

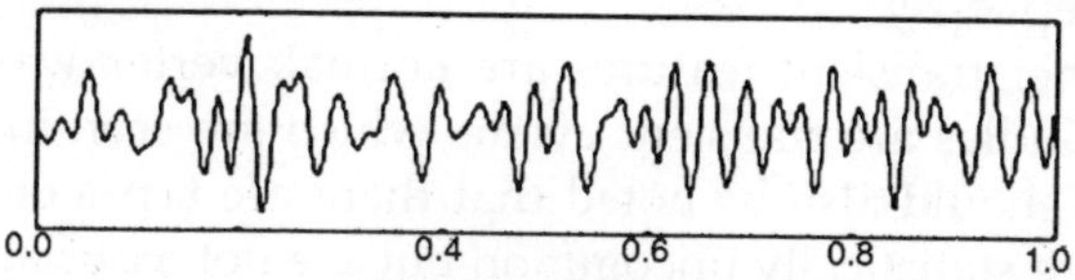

Fig. Beta Waves.

Low amplitude beta with multiple and varying frequencies is often associated with active, busy or anxious thinking and active concentration. Rhythmic beta with a dominant set of frequencies is associated with various pathologies and drug effects, especially benzodiazepines.

Activity over about 25 Hz seen in the scalp EEG is rarely cerebral (i.e., it is most often artifactual). It may be absent or reduced in areas of cortical damage. It is the dominant rhythm in patients who are alert or anxious or who have their eyes open.

Gamma Frequency

Gamma is the frequency range approximately 26–100 Hz. Because of the filtering properties of the skull and scalp, gamma rhythms can only be recorded from electro-corticography or possibly with magnetoen-cephalography.

Gamma rhythms are thought to represent binding of different populations of neurons together into a network for the purpose of carrying out a certain cognitive or motor function.

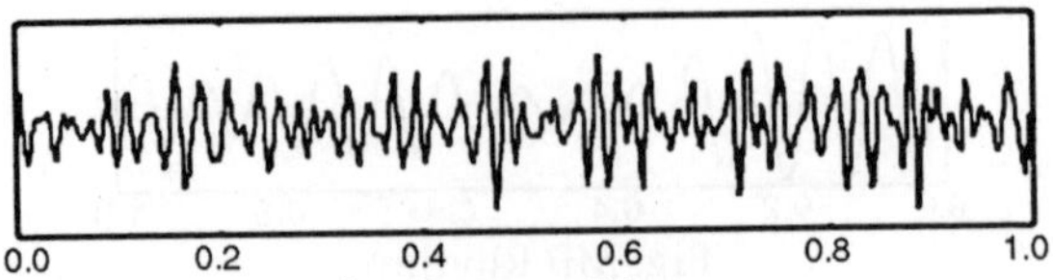

Fig. Gamma Waves.

"Ultra-slow" or "near-DC" activity is recorded using DC amplifiers in some research contexts. It is not typically recorded in a clinical context because the signal at these frequencies is susceptible to a number of artifacts. Some features of the EEG are transient rather than rhythmic. Spikes and sharp waves may represent seizure activity or interictal activity in individuals with epilepsy or a predisposition toward epilepsy.

Other transient features are normal: vertex waves and sleep spindles are transient events which are seen in normal sleep. It should also be noted that there are types of activity which are statistically uncommon but are not associated with dysfunction or disease. These are often referred to as "normal variants." The mu rhythm is an example of a normal variant.The normal EEG varies by age. The neonatal EEG is quite different from the adult EEG. The EEG in childhood is generally comprised of slower frequency oscillations than the adult EEG.

The normal EEG also varies depending on state. The EEG is used along with other measurements (EOG, EMG) to define sleep stages in polysomnography. Stage I sleep (equivalent to drowsiness in some systems) appears on the EEG as drop-out of the posterior basic rhythm. There can be an increase in theta frequencies.

Santamaria and Chiappa cataloged a number of the variety of patterns associated with drowsiness. Stage II sleep is characterized by sleep spindles—transient runs of rhythmic activity in the 12-14 Hz range (sometimes referred to as the "sigma" band) that have a frontal-central maximum.

Most of the activity in Stage II is in the 3-6 Hz range. Stage III and IV sleep are defined by the presence of delta frequences and are often referred to collectively as "slow-wave sleep." Stages I-IV are comprise non-REM (or "NREM") sleep.

The EEG in REM (rapid eye movement) sleep appears somewhat similar to the awake EEG. EEG under general anesthesia depends on the type of anesthetic employed. With halogenated anesthetics, such as halothane or intravenous agents, such as propofol, a rapid (alpha or low beta), nonreactive EEG pattern is seen over most of the scalp, especially anteriorly; in some older terminology this was known as a WAR (widespread anterior rapid) pattern, contrasted with a WAIS (widespread slow) pattern associated with high doses of opiates. Anesthetic effects on EEG signals are beginning to be understood at the level of drug actions on different kinds of synapses and the circuits that allow synchronized neuronal activity.

ARTIFACTS

Biological Artifacts

Signals in the EEG that are of non-cerebral origin are called artifacts. The EEG is nearly always contaminated by such signals. This is one of the reasons why it takes considerable experience to interpret EEGs clinically. The most common types of artifacts are:

- Eye artifacts
- EKG artifacts
- EMG artifacts
- Glossokinetic artifacts

Eyeball artifacts are caused by the potential difference between the cornea and retina, which is quite large compared to cerebral potentials. When the eye is completely still, this is not a problem. But there are nearly always small or large reflexive eye movements, which generates a potential which is picked up in the frontopolar and frontal leads. Eye movements - whether vertical or horizontal [saccades] - are caused by ocular muscles, which also generate electromyographic potentials.

Purposeful or reflexive eye blinking also generates electromyographic potentials, but more importantly there is reflexive movement of the eyeball during blinking which gives a characteristic artefactual appearance of the EEG. Eyelid fluttering artifacts of a characteristic type were previously called Kappa rhythm (or Kappa waves). It is usually seen in the prefrontal leads, that is, just over the eyes. Sometimes they are seen with mental activity.

They are usually in the Theta (4–7 Hz) or Alpha (8–13 Hz) range. They were named because they were believed to originate from the brain. Later study revealed they were generated by rapid fluttering of the eyelids, sometimes so minute that it was difficult to see. They are in fact noise or "artifact" in the EEG reading, and should not technically be called a rhythm or wave. Therefore, the term Kappa rhythm (or wave) is no longer used in electroencephalography.

It should be described as eyelid fluttering artifact. Some of these artifacts are useful. Eye movements are very important in polysomnography, and is also useful in conventional EEG for assessing possible changes in alertness, drowsiness or sleep. EKG artifacts are quite common and can be mistaken for spike activity. Because of this, modern EEG acquisition commonly includes a one-channel EKG from the extremities.

This also allows the EEG to identify cardiac arrythmias that are an important differential diagnosis to syncope or other episodic/attack disorders.

Glossokinetic artifacts are caused by the potential difference between the base and the tip of the tongue. Minor tongue movements can contaminate the EEG, especially in parkinsonian and tremor disorders.

Artifact Correction

Recently, source decomposition techniques have been used to correct or remove EEG contaminates. These techniques attempt to "unmix" the EEG signals into some number of underlying components. There are many source separation algorithms, often assuming various behaviours or natures of EEG. Regardless, the principle behind any particular method usually allow "remixing" only those components that would result in "clean" EEG by nullifying (zeroing) the weight of unwanted components.

External Artifacts

In addition to internal artifacts, there are many artifacts which originate from outside the patient. Movement by the patient, or even just settling of the electrodes, may cause *electrode pops,* spikes originating from a momentary change in the impedance of a given electrode.

Poor grounding of the EEG electrodes can cause significant 50 or 60Hz artifact, depending on the local power system's frequency.

A third source of possible interference can be the presence of an IV drip; such devices can cause rhythmic, fast, low-voltage bursts, which may be confused for spikes.

ABNORMAL ACTIVITY

Abnormal activity can broadly be separated into epileptiform and non-epileptiform activity. It can also be separated into focal or diffuse. Focal epileptiform discharges represent fast, synchronous potentials in a large number of neurons in a somewhat discrete area of the brain. These can occur as inter-ictal activity, between seizures, and represent an area of cortical irritability that may be predisposed to producing epileptic seizures.

Inter-ictal discharges are not wholly reliable for determining whether a patient has epilepsy nor where his/her seizure might originate. Generalized epileptiform discharges often have an anterior maximum, but the are seen synchronously throughout the entire brain. They are strongly suggestive of a generalized epilepsy.

Focal non-epileptiform abnormal activity may occur over areas of the brain where there is focal damage of the cortex or white matter.

It often consists of an increase in slow frequency rhythms and/or a loss of normal higher frequency rhythms.

It may also appear as focal or unilateral decrease in amplitude of the EEG signal. Diffuse non-epileptiform abnormal activity may manifest as diffuse abnormally slow rhythms or bilateral slowing of normal rhythms, such as the PBR.

Various Uses

The EEG has been used for many purposes besides the conventional uses of clinical diagnosis and conventional cognitive neuroscience.

Neurofeedback remains an important extension, and in its most advanced form is also attemped as the basis of brain computer interfaces.

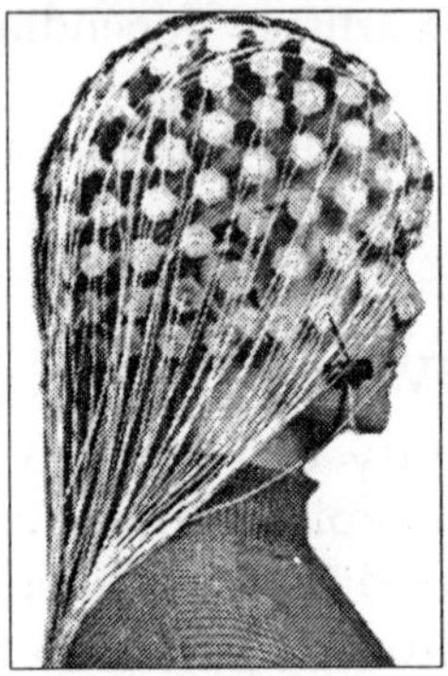

Fig. Girl Wearing Electrodes for Electroencephalography

There are many commercial products substantially based on the EEG.

Chapter 7

Electrocardiogram

An electrocardiogram (ECG or EKG, abbreviated from the German *Elektrokardiogramm*) is a graphic produced by an electrocardiograph, which records the electrical activity of the heart over time. Its name is made of different parts: *electro*, because it is related to electronics, *cardio*, Greek for heart, *gram*, a Greek root meaning "to write".

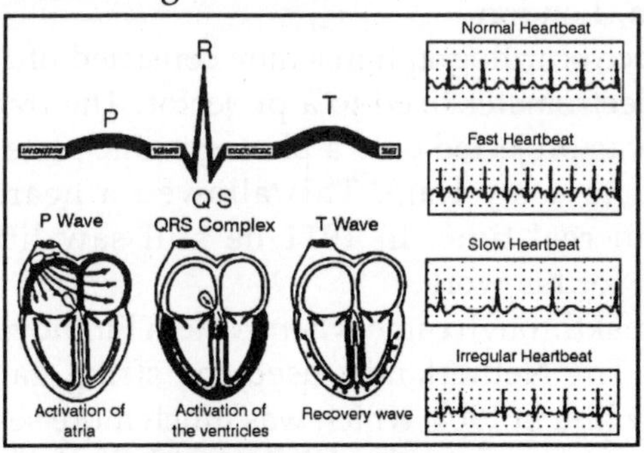

Fig. Electrocardiogram

The heart muscles create electrical waves when they pump. These waves pass through the body and can be measured at electrodes (electrical contacts) attached to the skin. Electrodes on different sides of the heart measure the activity of different muscles. An ECG displays the voltage between pairs of these electrodes, and the muscle activity that they measure from different directions. This display indicates the overall rhythm of the heart, and weaknesses in different muscles.

It is the best way to measure and diagnose abnormal rhythms of the heart, particularly abnormal rhythms caused by damage to the conductive tissue that carries electrical signals, or abnormal rhythms caused by levels of salts, such as calcium, that are too high or low.

In myocardial infarction (MI), the ECG can identify damaged heart muscle. But it can only identify damage to muscle in certain areas, so it can't rule out damage in other areas. The ECG cannot reliably measure the pumping ability of the heart; ultrasound is used for that.

Alexander Muirhead attached wires to a feverish patient's wrist to obtain a record of the patient's heartbeat while studying for his DSc (in electricity) in 1872 at St Bartholomew's Hospital. This activity was directly recorded and visualized using a Lippmann capillary electrometer by the British physiologist John Burdon Sanderson. The first to systematically approach the heart from an electrical point-of-view was Augustus Waller, working in St Mary's Hospital in Paddington, London.

His electrocardiograph machine consisted of a Lippmann capillary electrometer fixed to a projector. The trace from the heartbeat was projected onto a photographic plate which was itself fixed to a toy train. This allowed a heartbeat to be recorded in real time. In 1911 he still saw little clinical application for his work.

The breakthrough came when Willem Einthoven, working in Leiden, The Netherlands, used the string galvanometer invented by him in 1901, which was much more sensitive than the capillary electrometer that Waller used. Einthoven assigned the letters P, Q, R, S and T to the various deflections, and described the electrocardiographic features of a number of cardiovascular disorders. In 1924, he was awarded the Nobel Prize in Medicine for his discovery.

Though the basic principles of that era are still in use today, there have been many advances in electrocardiography over the years.

The instrumentation, has evolved from a cumbersome laboratory apparatus to compact electronic systems that often include computerized interpretation of the electrocardiogram.

ECG Graph Paper

A typical electrocardiograph runs at a paper speed of 25 mm/s, although faster paper speeds are occasionally used. Each small block of ECG paper is 1 mm^2. At a paper speed of 25 mm/s, one small block of ECG paper translates into 0.04 s (or 40 ms). Five small blocks make up 1 large block, which translates into 0.20 s (or 200 ms).

Hence, there are 5 large blocks per second. A diagnostic quality 12 lead ECG is calibrated at 10 mm/mV, so 1 mm translates into 0.1 mV. A "Calibration" signal should be included with every record. A standard signal of 1mV must move the stylus vertically 1 cm, that is two large squares on ECG paper.

Filter Selection

Modern ECG monitors offer multiple filters for signal processing. The most common settings are monitor mode and diagnostic mode. In monitor mode, the low frequency filter is set at either 0.5 Hz or 1 Hz and the high frequency filter is set at 40 Hz. This limits artifact for routine cardiac rhythm monitoring.

The high-pass filter helps reduce wandering baseline and the low pass filter helps reduce 50 or 60 Hz power line noise (the power line network frequency differs between 50 and 60 Hz in different countries). In diagnostic mode, the high pass filter is set at 0.05 Hz, which allows accurate ST segments to be recorded.

The low pass filter is set to 40, 100, or 150 Hz. Consequently, the monitor mode ECG display is more filtered than diagnostic mode, because its bandpass is narrower.

Leads

Graphic showing the relationship between positive electrodes, depolarization wavefronts (or mean electrical vectors), and complexes displayed on the ECG.

The word *lead* has two meanings in electrocardiography: it refers to either the wire that connects an electrode to the electrocardiograph, or (more commonly) to a combination of electrodes that form an imaginary line in the body along which the electrical signals are measured.

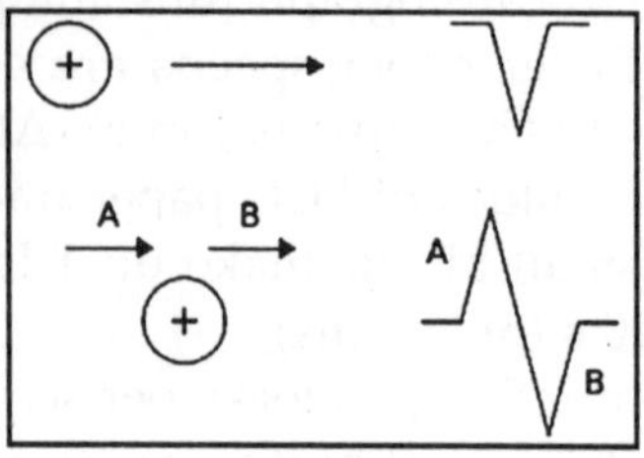

Thus, the term *loose lead artifact* uses the former meaning, while the term *12 lead ECG* uses the latter. In fact, a 12 lead electrocardiograph usually only uses 10 wires/electrodes. The latter definition of lead is the one used here.

An electrocardiogram is obtained by measuring electrical potential between various points of the body using a biomedical instrumentation amplifier.

A lead records the electrical signals of the heart from a particular combination of recording electrodes which are placed at specific points on the patient's body.

- When a depolarization wavefront (or mean electrical vector) moves toward a positive electrode, it creates a *positive* deflection on the ECG in the corresponding lead.
- When a depolarization wavefront (or mean electrical vector) moves away from a positive electrode, it creates a *negative* deflection on the ECG in the corresponding lead.
- When a depolarization wavefront (or mean electrical vector) moves perpendicular to a positive electrode, it creates an *equiphasic* (or isoelectric) complex on the ECG. It will be positive as the depolarization wavefront (or mean electrical vector) approaches (A), and then become negative as it passes by (B).

There are two types of leads—*unipolar* and *bipolar*. The former have an indifferent electrode at the centre of the Einthoven's triangle (which can be likened to a 'neutral' of the wall socket) at zero potential.

The direction of these leads is from the "centre" of the

heart radially outward and includes the precordial (chest) leads and limb leads— VL, VR, & VF.

The latter, in contrast, have both the electrodes at some potential and the direction of the corresponding electrode is from the electrode at lower potential to the one at higher potential, e.g., in limb lead I, the direction is from left to right. These include the limb leads—I, II, and III.

Note that the colouring scheme for leads varies by country.

LIMB

AUGMENTED LIMB

Leads aVR, aVL, and aVF are augmented limb leads. They are derived from the same three electrodes as leads I, II, and III. However, they view the heart from different angles (or vectors) because the negative electrode for these leads is a modification of Wilson's central terminal, which is derived by adding leads I, II, and III together and plugging them into the negative terminal of the EKG machine.

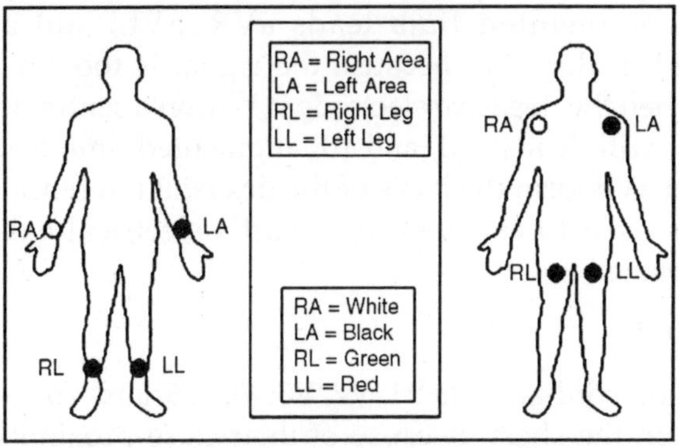

Fig. Proper Placement of the Limb Leads.

This zeroes out the negative electrode and allows the positive electrode to become the "exploring electrode" or a unipolar lead. This is possible because Einthoven's Law states that I + (-II) + III = 0. The equation can also be written I + III = II. It is written this way (instead of I + II + III = 0) because Einthoven reversed the polarity of lead II in Einthoven's triangle, possibly because he liked to view upright QRS complexes.

Wilson's central terminal paved the way for the development of the augmented limb leads aVR, aVL, aVF and the precordial leads V1, V2, V3, V4, V5, and V6.

Lead aVR or "augmented vector right" has the positive electrode (white) on the right arm. The negative electrode is a combination of the left arm (black) electrode and the left leg (red) electrode, which "augments" the signal strength of the positive electrode on the right arm.

Lead aVL or "augmented vector left" has the positive (black) electrode on the left arm. The negative electrode is a combination of the right arm (white) electrode and the left leg (red) electrode, which "augments" the signal strength of the positive electrode on the left arm.

- Lead aVF or "augmented vector foot" has the positive (red) electrode on the left leg. The negative electrode is a combination of the right arm (white) electrode and the left arm (black) electrode, which "augments" the signal of the positive electrode on the left leg.

The augmented limb leads aVR, aVL, and aVF are amplified in this way because the signal is too small to be useful when the negative electrode is Wilson's central terminal. Together with leads I, II, and III, augmented limb leads aVR, aVL, and aVF form the basis of the hexaxial reference system, which is used to calculate the heart's electrical axis in the frontal plane.

Precordial

The precordial leads V1, V2, V3, V4, V5, and V6 are placed directly on the chest. Because of their close proximity to the heart, they do not require augmentation.

Wilson's central terminal is used for the negative electrode, and these leads are considered to be unipolar. The precordial leads view the heart's electrical activity in the so-called horizontal plane.

The heart's electrical axis in the horizontal plane is referred to as the Z axis. Leads V1, V2, and V3 are referred to as the right precordial leads and V4, V5, and V6 are referred to as the left precordial leads.

The QRS complex should be negative in lead V1 and positive in lead V6. The QRS complex should show a gradual transition from negative to positive between leads V2 and V4.

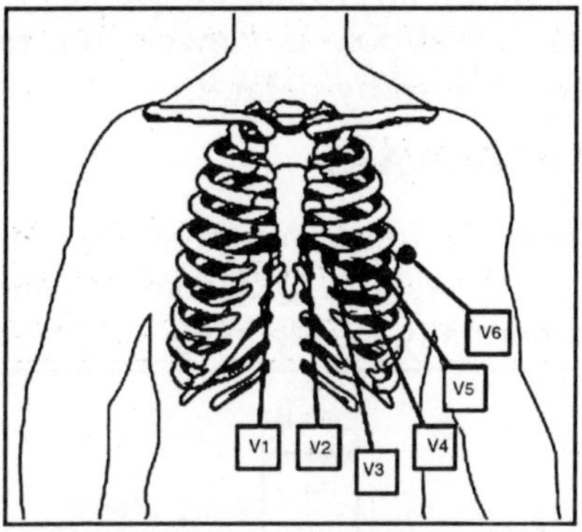

Fig. Proper Placement of the Precordial Leads.

The equiphasic lead is referred to as the transition lead. When the transition occurs earlier than lead V3, it is referred to as an early transition.

When it occurs later than lead V3, it is referred to as a late transition. There should also be a gradual increase in the amplitude of the R wave between leads V1 and V4. This is known as R wave progression. Poor R wave progression is a nonspecific finding.

It can be caused by conduction abnormalities, myocardial infarction, cardiomyopathy, and other pathological conditions.

- Lead V1 is placed in the fourth intercostal space to the right of the sternum.
- Lead V2 is placed in the fourth intercostal space to the left of the sternum.
- Lead V3 is placed directly between leads V2 and V4.
- Lead V4 is placed in the fifth intercostal space in the midclavicular line (even if the apex beat is displaced).
- Lead V5 is placed horizontally with V4 in the anterior axillary line
- Lead V6 is placed horizontally with V4 and V5 in the midaxillary line.

Ground

An additional electrode (usually green) is present in modern four-lead and twelve-lead ECGs. This is the ground lead and is placed on the right leg by convention, although in theory it can be placed anywhere on the body. With a three-lead ECG, when one dipole is viewed, the remaining lead becomes the ground lead by default.

WAVES AND INTERVALS

A typical ECG tracing of a normal heartbeat (or cardiac cycle) consists of a P wave, a QRS complex and a T wave. A small *U wave* is normally visible in 50 to 75% of ECGs.

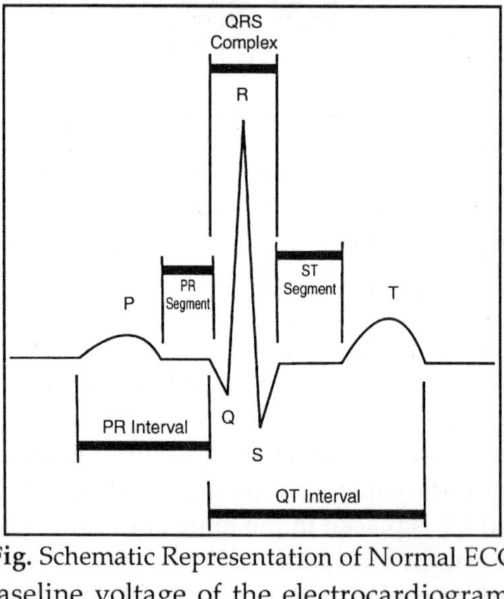

Fig. Schematic Representation of Normal ECG

The baseline voltage of the electrocardiogram is known as the isoelectric line. Typically the isoelectric line is measured as the portion of the tracing following the T wave and preceding the next P wave.

Rhythm Analysis

There are some basic rules that can be followed to identify a patient's heart rhythm. What is the rate? Is it regular or irregular?

Are P waves present? Are QRS complexes present? Is there a 1:1 ratio between P waves and QRS complexes? Is the PR interval constant?

P wave

During normal atrial depolarization, the main electrical vector is directed from the SA node towards the AV node, and spreads from the right atrium to the left atrium.

This turns into the P wave on the ECG, which is upright in II, III, and aVF (since the general electrical activity is going toward the positive electrode in those leads), and inverted in aVR (since it is going away from the positive electrode for that lead).

A P wave must be upright in leads II and aVF and inverted in lead aVR to designate a cardiac rhythm as Sinus Rhythm.

- The relationship between P waves and QRS complexes helps distinguish various cardiac arrhythmias.
- The shape and duration of the P waves may indicate atrial enlargement.

PR Interval

The PR interval is measured from the beginning of the P wave to the beginning of the QRS complex. It is usually 120 to 200 ms long. On an ECG tracing, this corresponds to 3 to 5 small boxes.

- A PR interval of over 200 ms may indicate a first degree heart block.
- A short PR interval may indicate a pre-excitation syndrome via an accessory pathway that leads to early activation of the ventricles, such as seen in Wolff-Parkinson-White syndrome.
- A variable PR interval may indicate other types of heart block.
- PR segment depression may indicate atrial injury or pericarditis.
- Variable morphologies of P waves in a single ECG lead is suggestive of an ectopic pacemaker rhythm such as wandering pacemaker or multifocal atrial tachycardia

QRS Complex

The QRS complex is a structure on the ECG that corresponds to the depolarization of the ventricles. Because the ventricles contain more muscle mass than the atria, the QRS complex is larger than the P wave.

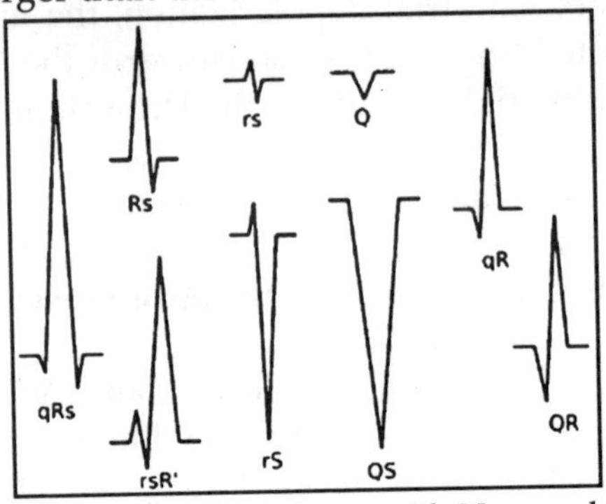

Fig. Various QRS Complexes with Nomenclature.

In addition, because the His/Purkinje system coordinates the depolarization of the ventricles, the QRS complex tends to look "spiked" rather than rounded due to the increase in conduction velocity.

A normal QRS complex is 0.06 to 0.10 sec (60 to 100 ms) in duration represented by three small squares or less, but any abnormality of conduction takes longer, and causes widened QRS complexes. Not every QRS complex contains a Q wave, an R wave, and an S wave. By convention, any combination of these waves can be referred to as a QRS complex. However, correct interpretation of difficult ECGs requires exact labeling of the various waves.

Some authors use lowercase and capital letters, depending on the relative size of each wave. An Rs complex would be positively deflected, while a rS complex would be negatively deflected. If both complexes were labeled RS, it would be impossible to appreciate this distinction without viewing the actual ECG.

- The duration, amplitude, and morphology of the QRS complex is useful in diagnosing cardiac arrhythmias, conduction abnormalities, ventricular hypertrophy, myocardial infarction, electrolyte derangements, and other disease states.

- Q waves can be normal (physiological) or pathological. Normal Q waves, when present, represent depolarization of the interventricular septum. For this reason, they are referred to as septal Q waves, and can be appreciated in the lateral leads I, aVL, V5 and V6.
- Q waves greater than 1/3 the height of the R wave, greater than 0.04 sec (40 ms) in duration, or in the right precordial leads are considered to be abnormal, and may represent myocardial infarction.

ST Segment

The ST segment connects the QRS complex and the T wave and has a duration of 0.08 to 0.12 sec (80 to 120 ms). It starts at the J point (junction between the QRS complex and ST segment) and ends at the beginning of the T wave. However, since it is usually difficult to determine exactly where the ST segment ends and the T wave begins, the relationship between the ST segment and T wave should be examined together. The typical ST segment duration is usually around 0.08 sec (80 ms). It should be essentially level with the PR and TP segment.

- The normal ST segment has a slight upward concavity.
- Flat, downsloping, or depressed ST segments may indicate coronary ischemia.
- ST segment elevation may indicate myocardial infarction. An elevation of >1mm and longer than 80 milliseconds following the J-point. This measure has a false positive rate of 15-20% (which is slightly higher in women than men) and a false negative rate of 20-30%.

T Wave

The T wave represents the repolarization (or recovery) of the ventricles. The interval from the beginning of the QRS complex to the apex of the T wave is referred to as the absolute refractory period. The last half of the T wave is referred to as the relative refractory period (or vulnerable period).

In most leads, the T wave is positive. However, a negative T wave is normal in lead aVR. Lead V1 may have a positive, negative, or biphasic T wave. In addition, it is not uncommon to have an isolated negative T wave in lead III, aVL, or aVF.

- Inverted (or negative) T waves can be a sign of coronary ischemia, Wellens' syndrome, left ventricular hypertrophy, or CNS disorder.
- Tall or "tented" symmetrical T waves may indicate hyperkalemia. Flat T waves may indicate coronary ischemia or hypokalemia.
- The earliest electrocardiographic finding of acute myocardial infarction is sometimes the hyperacute T wave, which can be distinguished from hyperkalemia by the broad base and slight asymmetry.
- When a conduction abnormality (e.g., bundle branch block, paced rhythm) is present, the T wave should be deflected opposite the terminal deflection of the QRS complex. This is known as appropriate T wave discordance.

QT Interval

The QT interval is measured from the beginning of the QRS complex to the end of the T wave. Normal values for the QT interval are between 0.30 and 0.44 (0.45 for women) seconds. The QT interval as well as the corrected QT interval are important in the diagnosis of long QT syndrome and short QT syndrome. The QT interval varies based on the heart rate, and various correction factors have been developed to correct the QT interval for the heart rate.

The most commonly used method for correcting the QT interval for rate is the one formulated by Bazett and published in 1920. Bazett's formula is $QT_c = \frac{QT}{\sqrt{RR}}$, where QTc is the QT interval corrected for rate, and RR is the interval from the onset of one QRS complex to the onset of the next QRS complex, measured in seconds. However, this formula tends to be inaccurate, and over-corrects at high heart rates and under-corrects at low heart rates.

U wave

The U wave is not always seen. It is typically small, and, by definition, follows the T wave. U waves are thought to represent repolarization of the papillary muscles or Purkinje fibres.

Prominent U waves are most often seen in hypokalemia, but may be present in hypercalcemia, thyrotoxicosis, or exposure to digitalis, epinephrine, and Class 1A and 3 antiarrhythmics, as well as in congenital long QT syndrome and in the setting of intracranial hemorrhage. An inverted U wave may represent myocardial ischemia or left ventricular volume overload.

CLINICAL LEAD GROUPS

There are twelve leads in total, each recording the electrical activity of the heart from a different perspective, which also correlate to different anatomical areas of the heart for the purpose of identifying acute coronary ischemia or injury. Two leads that look at the same anatomical area of the heart are said to be *contiguous*.

- The inferior leads (leads II, III and aVF) look at electrical activity from the vantage point of the inferior (or diaphragmatic) wall of the left ventricle.
- The lateral leads (I, aVL, V_5 and V_6) look at the electrical activity from the vantage point of the lateral wall of left ventricle. Because the positive electrode for leads I and aVL are located on the left shoulder, leads I and aVL are sometimes referred to as the high lateral leads. Because the positive electrodes for leads V_5 and V_6 are on the patient's chest, they are sometimes referred to as the low lateral leads.
- The septal leads, V_1 and V_2 look at electrical activity from the vantage point of the septal wall of the left ventricle. They are often grouped together with the anterior leads.
- The anterior leads, V_3 and V_4 look at electrical activity from the vantage point of the anterior wall of the left ventricle.

- In addition, any two precordial leads that are next to one another are considered to be contiguous. Even though V4 is an anterior lead and V5 is a lateral lead, they are contiguous because they are next to one another.
- Lead aVR offers no specific view of the left ventricle. Rather, it views the inside of the endocardial wall from its perspective on the right shoulder.

ELECTRICAL CONDUCTION SYSTEM OF THE HEART

The normal electrical conduction in the heart allows the impulse that is generated by the sinoatrial node (SA node) of the heart to be propagated to (and stimulate) the myocardium (Cardiac muscle).

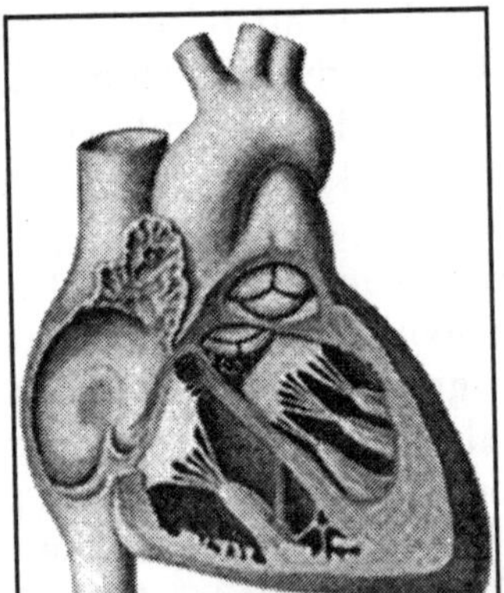

The myocardium contracts after stimulation. It is the ordered stimulation of the myocardium that allows efficient contraction of the heart, thereby allowing blood to be pumped throughout the body.

Requirements for Effective Pumping

In order to maximize efficiency of contraction and cardiac output, the conduction system of the heart has:

- Substantial atrial to ventricular delay. This allows the atria to completely empty their contents into the ventricles; simultaneous contraction would cause inefficient filling and backflow. The atria are electrically isolated from the ventricles, connected only via the AV node which briefly delays the signal.

- Coordinated contraction of ventricular cells. The ventricles must maximize systolic pressure to force blood through the circulation, so all the ventricular cells must work together.
  - Ventricular contraction begins at the apex of the heart, progressing upwards to eject blood into the great arteries. Contraction that squeezes blood towards the exit is more efficient than a simple squeeze from all directions. Although the ventricular stimulus originates from the AV node in the wall separating the atria and ventricles, the Bundle of His conducts the signal to the apex.
  - Depolarization propagates through cardiac muscle very rapidly. *Cells of the ventricles contract nearly simultaneously.*
  - The action potentials of cardiac muscle are unusually sustained. *This prevents premature relaxation, maintaining initial contraction until the entire myocardium has had time to depolarize and contract.*
- Absence of tetany. After contracting, the heart must relax to fill up again. Sustained contraction of the heart without relaxation would be fatal, and this is prevented by a temporary inactivation of certain ion channels.

Electrochemical Mechanism

Cardiac muscle has some similarities to neurons and skeletal muscle, as well as important unique properties. Like a neuron, a given myocardial cell has a negative membrane potential when at rest. Stimulation above a threshold value induces the opening of voltage-gated ion channels and a flood of cations into the cell. The positively charged ions entering the cell cause the depolarization characteristic of an action potential. Like skeletal muscle, depolarization causes the opening of voltage-gated calcium channels and release of Ca^{2+} from the t-tubules.

This influx of calcium causes calcium-induced calcium release from the sarcoplasmic reticulum, and free Ca^{2+} causes muscle contraction. After a delay (the absolute refractory period), Potassium channels reopen and the resulting flow of K^+ out of the cell causes repolarization to the resting state.

Note that there are important physiological differences between nodal cells and ventricular cells; the specific differences in ion channels and mechanisms of polarization give rise to unique properties of SA node cells, most importantly the spontaneous depolarizations necessary for the SA node's pacemaker activity.

Conduction Pathway

Signals arising in the SA node stimulate the atria to contract and travel to the AV node. After a delay, the stimulus is conducted through the bundle of His to the Purkinje fibres and the endocardium at the apex of the heart, then finally to the ventricular epicardium.

Microscopically, the wave of depolarization propagates to adjacent cells via gap junctions located on the intercalated disk. The heart is a *syncytium*: electrical impulses propagate freely between cells in every direction, so that the myocardium functions as a single contractile unit.

This property allows rapid, synchronous depolarization of the myocardium. While normally advantageous, this property can be detrimental as it potentially allows the propagation of incorrect electrical signals. These gap junctions can close to isolate damaged or dying tissue, as in a myocardial infarction.

DEPOLARIZATION AND THE ECG

SA node: P wave

Under normal conditions, electrical activity is spontaneously generated by the SA node, the physiological pacemaker.

This electrical impulse is propagated throughout the right atrium, and through Bachman's Bundle to the left atrium, stimulating the myocardium of the atria to contract. The conduction of the electrical impulse throughout the atria is seen on the ECG as the P wave. As the electrical activity is spreading throughout the atria, it travels via specialized pathways, known as *internodal tracts,* from the SA node to the AV node.

AV Node/Bundles: PR Interval

The AV node functions as a critical delay in the conduction system. Without this delay, the atria and ventricles would contract at the same time, and blood wouldn't flow effectively from the atria to the ventricles.

The delay in the AV node forms much of the PR segment on the ECG. And part of atrial repolarization can be represented by PR segment.

The distal portion of the AV node is known as the Bundle of His.

The Bundle of His splits into two branches in the interventricular septum, the left bundle branch and the right bundle branch.

The left bundle branch activates the left ventricle, while the right bundle branch activates the right ventricle. The left bundle branch is short, splitting into the left anterior fascicle and the left posterior fascicle.

The left posterior fascicle is relatively short and broad, with dual blood supply, making it particularly resistant to ischemic damage.

Pathology

An impulse (action potential) that originates from the SA node at a rate of 60 - 100 beats/minute (bpm) is known as *normal sinus rhythm.*

If SA nodal impulses occur at a rate less than 60 bpm, the heart rhythm is known as sinus bradycardia. If SA nodal impulse occur at a rate exceeding 100 bpm, the consequent rapid heart rate is sinus tachycardia.

These conditions are not necessarily bad symptoms, however. Trained athletes, usually show heart rates slower than 60bpm when not exercising.

If the SA node fails to initialize the AV Junction can take over as the main pacemaker of the heart. The AV Junction "surrounds" the AV node (the AV node is not able to initialize its own impulses) and has a regular rate of 40 to 60 bpm.

These "Junctional" rhythms are characterized by a missing or inverted P-Wave. If both the SA node and the AV Junction fail to initialize the electrical impulse, the ventricles can fire the electrical impulses themselves at a rate of 20 to 40 bpm and will have a QRS complex of greater than.12ms.

Chapter 8

Deoxyribo Nucleic Acid

DNA

The DNA segments that carry this genetic information are called genes, but other DNA sequences have structural purposes, or are involved in regulating the use of this genetic information. Deoxyribonucleic acid (DNA) is a nucleic acid that contains the genetic instructions used in the development and functioning of all known living organisms.

The main role of DNA molecules is the long-term storage of information. DNA is often compared to a set of blueprints, since it contains the instructions needed to construct other components of cells, such as proteins and RNA molecules. Chemically, DNA is a long polymer of simple units called nucleotides, with a backbone made of sugars and phosphate groups joined by ester bonds.

Attached to each sugar is one of four types of molecules called bases. It is the sequence of these four bases along the backbone that encodes information. This information is read using the genetic code, which specifies the sequence of the amino acids within proteins. The code is read by copying stretches of DNA into the related nucleic acid RNA, in a process called transcription. Within cells, DNA is organized into structures called chromosomes. These chromosomes are duplicated before cells divide, in a process called DNA replication.

Eukaryotic organisms such as animals, plants, and fungi store their DNA inside the cell nucleus, while in prokaryotes such as bacteria it is found in the cell's cytoplasm. Within the chromosomes, chromatin proteins such as histones compact and organize DNA.

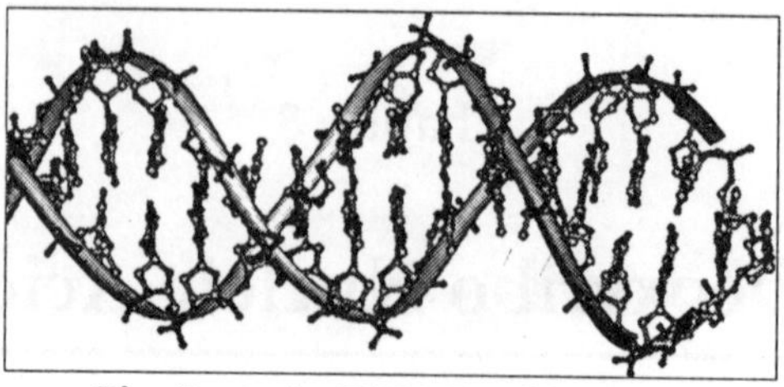

Fig. Part of a DNA Double Helix

These compact structures guide the interactions between DNA and other proteins, helping control which parts of the DNA are transcribed.

PHYSICAL AND CHEMICAL PROPERTIES

DNA is a long polymer made from repeating units called nucleotides. The DNA chain is 22 to 26 Ångströms wide (2.2 to 2.6 nanometres), and one nucleotide unit is 3.3 Å (0.33 nm) long. Although each individual repeating unit is very small, DNA polymers can be enormous molecules containing millions of nucleotides. For instance, the largest human chromosome, chromosome number 1, is 220 million base pairs long.

In living organisms, DNA does not usually exist as a single molecule, but instead as a tightly-associated pair of molecules. These two long strands entwine like vines, in the shape of a double helix. The nucleotide repeats contain both the segment of the backbone of the molecule, which holds the chain together, and a base, which interacts with the other DNA strand in the helix. In general, a base linked to a sugar is called a nucleoside and a base linked to a sugar and one or more phosphate groups is called a nucleotide. If multiple nucleotides are linked together, as in DNA, this polymer is called a polynucleotide.

The backbone of the DNA strand is made from alternating phosphate and sugar residues. The sugar in DNA is 2-deoxyribose, which is a pentose (five-carbon) sugar. The sugars are joined together by phosphate groups that form phosphodiester bonds between the third and fifth carbon atoms of adjacent sugar rings.

These asymmetric bonds mean a strand of DNA has a direction. In a double helix the direction of the nucleotides in one strand is opposite to their direction in the other strand. This arrangement of DNA strands is called antiparallel. The asymmetric ends of DNA strands are referred to as the 52 (*five prime*) and 32 (*three prime*) ends. One of the major differences between DNA and RNA is the sugar, with 2-deoxyribose being replaced by the alternative pentose sugar ribose in RNA.

The DNA double helix is stabilized by hydrogen bonds between the bases attached to the two strands. The four bases found in DNA are adenine (abbreviated A), cytosine (C), guanine (G) and thymine (T). These four bases are attached to the sugar/phosphate to form the complete nucleotide, as shown for adenosine monophosphate.

These bases are classified into two types; adenine and guanine are fused five- and six-membered heterocyclic compounds called purines, while cytosine and thymine are six-membered rings called pyrimidines. A fifth pyrimidine base, called uracil (U), usually takes the place of thymine in RNA and differs from thymine by lacking a methyl group on its ring. Uracil is not usually found in DNA, occurring only as a breakdown product of cytosine.

Sense and Antisense

A DNA sequence is called "sense" if its sequence is the same as that of a messenger RNA copy that is translated into protein. The sequence on the opposite strand is called the "antisense" sequence. Both sense and antisense sequences can exist on different parts of the same strand of DNA (i.e. both strands contain both sense and antisense sequences). In both prokaryotes and eukaryotes, antisense RNA sequences are produced, but the functions of these RNAs are not entirely clear.

One proposal is that antisense RNAs are involved in regulating gene expression through RNA-RNA base pairing. A few DNA sequences in prokaryotes and eukaryotes, and more in plasmids and viruses, blur the distinction made above between sense and antisense strands by having overlapping genes. In these cases, some DNA sequences do double duty, encoding one protein when read 52 to 32 along one strand, and a second protein when read in the opposite direction (still 52 to 32) along the other strand.

In bacteria, this overlap may be involved in the regulation of gene transcription, while in viruses, overlapping genes increase the amount of information that can be encoded within the small viral genome.

Supercoiling

DNA can be twisted like a rope in a process called DNA supercoiling. With DNA in its "relaxed" state, a strand usually circles the axis of the double helix once every 10.4 base pairs, but if the DNA is twisted the strands become more tightly or more loosely wound. If the DNA is twisted in the direction of the helix, this is positive supercoiling, and the bases are held more tightly together.

If they are twisted in the opposite direction, this is negative supercoiling, and the bases come apart more easily. In nature, most DNA has slight negative supercoiling that is introduced by enzymes called topoisomerases. These enzymes are also needed to relieve the twisting stresses introduced into DNA strands during processes such as transcription and DNA replication.

Alternative Double-helical Structures

DNA exists in many possible conformations. However, only A-DNA, B-DNA, and Z-DNA have been observed in organisms. Which conformation DNA adopts depends on the sequence of the DNA, the amount and direction of supercoiling, chemical modifications of the bases and also solution conditions, such as the concentration of metal ions and polyamines. Of these three conformations, the "B" form is most common under the conditions found in cells. The two alternative double-helical forms of DNA differ in their geometry and dimensions.

The A form is a wider right-handed spiral, with a shallow, wide minor groove and a narrower, deeper major groove. The A form occurs under non-physiological conditions in dehydrated samples of DNA, while in the cell it may be produced in hybrid pairings of DNA and RNA strands, as well as in enzyme-DNA complexes.

Segments of DNA where the bases have been chemically-modified by methylation may undergo a larger change in conformation and adopt the Z form. Here, the strands turn about the helical axis in a left-handed spiral, the opposite of the more common B form. These unusual structures can be recognized by specific Z-DNA binding proteins and may be involved in the regulation of transcription.

Quadruplex Structures

At the ends of the linear chromosomes are specialized regions of DNA called telomeres. The main function of these regions is to allow the cell to replicate chromosome ends using the enzyme telomerase, as the enzymes that normally replicate DNA cannot copy the extreme 32 ends of chromosomes. These specialized chromosome caps also help protect the DNA ends, and stop the DNA repair systems in the cell from treating them as damage to be corrected. In human cells, telomeres are usually lengths of single-stranded DNA containing several thousand repeats of a simple TTAGGG sequence.

These guanine-rich sequences may stabilize chromosome ends by forming structures of stacked sets of four-base units, rather than the usual base pairs found in other DNA molecules. Here, four guanine bases form a flat plate and these flat four-base units then stack on top of each other, to form a stable *G-quadruplex* structure.

These structures are stabilized by hydrogen bonding between the edges of the bases and chelation of a metal ion in the centre of each four-base unit. Other structures can also be formed, with the central set of four bases coming from either a single strand folded around the bases, or several different parallel strands, each contributing one base to the central structure.

In addition to these stacked structures, telomeres also form large loop structures called telomere loops, or T-loops. Here, the single-stranded DNA curls around in a long circle stabilized by telomere-binding proteins. At the very end of the T-loop, the single-stranded telomere DNA is held onto a region of double-stranded DNA by the telomere strand disrupting the double-helical DNA and base pairing to one of the two strands. This triple-stranded structure is called a displacement loop or D-loop.

Major and Minor Grooves

The double helix is a right-handed spiral. As the DNA strands wind around each other, they leave gaps between each set of phosphate backbones, revealing the sides of the bases inside. There are two of these grooves twisting around the surface of the double helix: one groove, the major groove, is 22 Å wide and the other, the minor groove, is 12 Å wide.

The narrowness of the minor groove means that the edges of the bases are more accessible in the major groove. As a result, proteins like transcription factors that can bind to specific sequences in double-stranded DNA usually make contacts to the sides of the bases exposed in the major groove.

Base Pairing

Each type of base on one strand forms a bond with just one type of base on the other strand. This is called complementary base pairing. Here, purines form hydrogen bonds to pyrimidines, with A bonding only to T, and C bonding only to G. This arrangement of two nucleotides binding together across the double helix is called a base pair. The double helix is also stabilized by the hydrophobic effect and pi stacking, which are not influenced by the sequence of the DNA. As hydrogen bonds are not covalent, they can be broken and rejoined relatively easily.

The two strands of DNA in a double helix can therefore be pulled apart like a zipper, either by a mechanical force or high temperature. As a result of this complementarity, all the information in the double-stranded sequence of a DNA helix is duplicated on each strand, which is vital in DNA replication. Indeed, this reversible and specific interaction between complementary base pairs is critical for all the functions of DNA in living organisms.

The two types of base pairs form different numbers of hydrogen bonds, AT forming two hydrogen bonds, and GC forming three hydrogen bonds. The GC base pair is therefore stronger than the AT base pair.

As a result, it is both the percentage of GC base pairs and the overall length of a DNA double helix that determine the strength of the association between the two strands of DNA. Long DNA helices with a high GC content have stronger-interacting strands, while short helices with high AT content have weaker-interacting strands. In biology, parts of the DNA double helix that need to separate easily, such as the TATAAT Pribnow box in some promoters, tend to have a high AT content, making the strands easier to pull apart.

In the laboratory, the strength of this interaction can be measured by finding the temperature required to break the hydrogen bonds, their melting temperature (also called T_m value). When all the base pairs in a DNA double helix melt, the strands separate and exist in solution as two entirely independent molecules. These single-stranded DNA molecules have no single common shape, but some conformations are more stable than others.

CHEMICAL MODIFICATIONS

Structure of cytosine with and without the 5-methyl group.

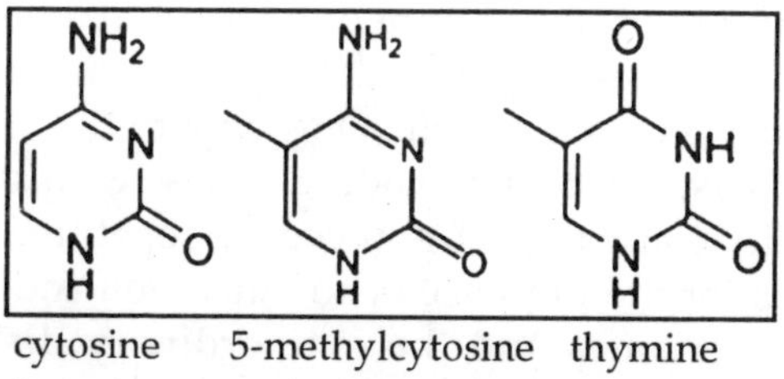

cytosine 5-methylcytosine thymine

After deamination the 5-methylcytosine has the same structure as thymine

Overview of Biological Functions

DNA usually occurs as linear chromosomes in eukaryotes, and circular chromosomes in prokaryotes. The set of chromosomes in a cell makes up its genome; the human genome has approximately 3 billion base pairs of DNA arranged into 46 chromosomes. The information carried by DNA is held in the sequence of pieces of DNA called genes.

Transmission of genetic information in genes is achieved via complementary base pairing. In transcription, when a cell uses the information in a gene, the DNA sequence is copied into a complementary RNA sequence through the attraction between the DNA and the correct RNA nucleotides. Usually, this RNA copy is then used to make a matching protein sequence in a process called translation which depends on the same interaction between RNA nucleotides. Alternatively, a cell may simply copy its genetic information in a process called DNA replication.

Genes and Genomes

Genomic DNA is located in the cell nucleus of eukaryotes, as well as small amounts in mitochondria and chloroplasts. In prokaryotes, the DNA is held within an irregularly shaped body in the cytoplasm called the nucleoid. The genetic information in a genome is held within genes, and the complete set of this information in an organism is called its genotype. A gene is a unit of heredity and is a region of DNA that influences a particular characteristic in an organism. Genes contain an open reading frame that can be transcribed, as well as regulatory sequences such as promoters and enhancers, which control the transcription of the open reading frame.

In many species, only a small fraction of the total sequence of the genome encodes protein. Only about 1.5% of the human genome consists of protein-coding exons, with over 50% of human DNA consisting of non-coding repetitive sequences. The reasons for the presence of so much non-coding DNA in eukaryotic genomes and the extraordinary differences in genome size, or *C-value*, among species represent a long-standing puzzle known as the "C-value enigma." However,

DNA sequences that do not code protein may still encode functional non-coding RNA molecules, which are involved in the regulation of gene expression.

Some non-coding DNA sequences play structural roles in chromosomes. Telomeres and centromeres typically contain few genes, but are important for the function and stability of chromosomes. An abundant form of non-coding DNA in humans are pseudogenes, which are copies of genes that have been disabled by mutation. These sequences are usually just molecular fossils, although they can occasionally serve as raw genetic material for the creation of new genes through the process of gene duplication and divergence.

Transcription and Translation

A gene is a sequence of DNA that contains genetic information and can influence the phenotype of an organism. Within a gene, the sequence of bases along a DNA strand defines a messenger RNA sequence, which then defines one or more protein sequences. The relationship between the nucleotide sequences of genes and the amino-acid sequences of proteins is determined by the rules of translation, known collectively as the genetic code. The genetic code consists of three-letter 'words' called *codons* formed from a sequence of three nucleotides (e.g. ACT, CAG, TTT).

In transcription, the codons of a gene are copied into messenger RNA by RNA polymerase. This RNA copy is then decoded by a ribosome that reads the RNA sequence by base-pairing the messenger RNA to transfer RNA, which carries amino acids. Since there are 4 bases in 3-letter combinations, there are 64 possible codons (4^3 combinations). These encode the twenty standard amino acids, giving most amino acids more than one possible codon. There are also three 'stop' or 'nonsense' codons signifying the end of the coding region; these are the TAA, TGA and TAG codons.

Base Modifications

The expression of genes is influenced by the chromatin structure of a chromosome and regions of that have low or no gene expression usually contain high levels of methylation of cytosine bases.

Cytosine methylation, producing 5-methylcytosine, is important for X-chromosome inactivation. The average level of methylation varies between organisms, with *Caenorhabditis elegans* lacking cytosine methylation, while vertebrates show higher levels, with up to 1% of their DNA containing 5-methylcytosine.

Despite the biological role of 5-methylcytosine it can deaminate to leave a thymine base, methylated cytosines are therefore particularly prone to mutations. Other base modifications include adenine methylation in bacteria and the glycosylation of uracil to produce the "J-base" in kinetoplastids.

DNA Damage

DNA can be damaged by many different sorts of mutagens, which are agents that change the DNA sequence. These agents include oxidizing agents, alkylating agents and also high-energy electromagnetic radiation such as ultraviolet light and X-rays. The type of DNA damage produced depends on the type of mutagen. UV light mostly damages DNA by producing thymine dimers, which are cross-links between adjacent pyrimidine bases in a DNA strand.

On the other hand, oxidants such as free radicals or hydrogen peroxide produce multiple forms of damage, including base modifications, particularly of guanosine, as well as double-strand breaks. It has been estimated that in each human cell, about 500 bases suffer oxidative damage per day. Of these oxidative lesions, the most dangerous are double-strand breaks, as these are difficult to repair and can produce point mutations, insertions and deletions from the DNA sequence, as well as chromosomal translocations. Many mutagens intercalate into the space between two adjacent base pairs.

Intercalators are mostly aromatic and planar molecules, and include ethidium, daunomycin, doxorubicin and thalidomide. In order for an intercalator to fit between base pairs, the bases must separate, distorting the DNA strands by unwinding of the double helix. These structural changes inhibit both transcription and DNA replication, causing toxicity and mutations.

As a result, DNA intercalators are often carcinogens, with benzopyrene diol epoxide, acridines, aflatoxin and ethidium bromide being well-known examples.

Nevertheless, due to their properties of inhibiting DNA transcription and replication, they are also used in chemotherapy to inhibit rapidly-growing cancer cells.

Replication

Cell division is essential for an organism to grow, but when a cell divides it must replicate the DNA in its genome so that the two daughter cells have the same genetic information as their parent.

The double-stranded structure of DNA provides a simple mechanism for DNA replication. Here, the two strands are separated and then each strand's complementary DNA sequence is recreated by an enzyme called DNA polymerase.

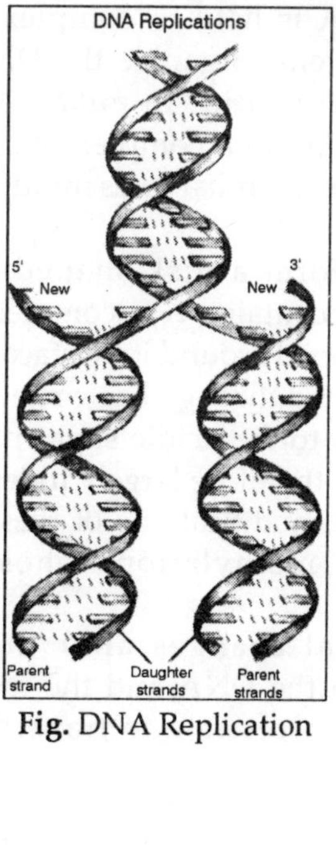

Fig. DNA Replication

This enzyme makes the complementary strand by finding the correct base through complementary base pairing, and bonding it onto the original strand. As DNA polymerases can only extend a DNA strand in a 5' to 3' direction, different mechanisms are used to copy the antiparallel strands of the double helix. In this way, the base on the old strand dictates which base appears on the new strand, and the cell ends up with a perfect copy of its DNA.

Interactions with Proteins

All the functions of DNA depend on interactions with proteins. These protein interactions can be non-specific, or the protein can bind specifically to a single DNA sequence. Enzymes can also bind to DNA and of these, the polymerases that copy the DNA base sequence in transcription and DNA replication are particularly important.

DNA-binding Proteins

Structural proteins that bind DNA are well-understood examples of non-specific DNA-protein interactions. Within chromosomes, DNA is held in complexes with structural proteins. These proteins organize the DNA into a compact structure called chromatin. In eukaryotes this structure involves DNA binding to a complex of small basic proteins called histones, while in prokaryotes multiple types of proteins are involved.

The histones form a disk-shaped complex called a nucleosome, which contains two complete turns of double-stranded DNA wrapped around its surface. These non-specific interactions are formed through basic residues in the histones making ionic bonds to the acidic sugar-phosphate backbone of the DNA, and are therefore largely independent of the base sequence. Chemical modifications of these basic amino acid residues include methylation, phosphorylation and acetylation.

These chemical changes alter the strength of the interaction between the DNA and the histones, making the DNA more or less accessible to transcription factors and changing the rate of transcription. Other non-specific DNA-binding proteins found in chromatin include the high-mobility group proteins, which bind preferentially to bent or distorted DNA. These proteins are important in bending arrays of nucleosomes and arranging them into more complex chromatin structures. A distinct group of DNA-binding proteins are the single-stranded-DNA-binding proteins that specifically bind single-stranded DNA.

In humans, replication protein A is the best-characterised member of this family and is essential for most processes where the double helix is separated, including DNA replication, recombination and DNA repair. These binding proteins seem to stabilize single-stranded DNA and protect it from forming stem-loops or being degraded by nucleases. In contrast, other proteins have evolved to specifically bind particular DNA sequences. The most intensively studied of these are the various classes of transcription factors, which are proteins that regulate transcription.

Each one of these proteins bind to one particular set of DNA sequences and thereby activates or inhibits the transcription of genes with these sequences close to their promoters. The transcription factors do this in two ways. Firstly, they can bind the RNA polymerase responsible for transcription, either directly or through other mediator proteins; this locates the polymerase at the promoter and allows it to begin transcription.

Alternatively, transcription factors can bind enzymes that modify the histones at the promoter; this will change the accessibility of the DNA template to the polymerase. As these DNA targets can occur throughout an organism's genome, changes in the activity of one type of transcription factor can affect thousands of genes.

Consequently, these proteins are often the targets of the signal transduction processes that mediate responses to environmental changes or cellular differentiation and development. The specificity of these transcription factors' interactions with DNA come from the proteins making multiple contacts to the edges of the DNA bases, allowing them to "read" the DNA sequence. Most of these base-interactions are made in the major groove, where the bases are most accessible.

DNA-MODIFYING ENZYMES

GENETIC RECOMBINATION

A DNA helix usually does not interact with other segments of DNA, and in human cells the different chromosomes even occupy separate areas in the nucleus called "chromosome territories".

This physical separation of different chromosomes is important for the ability of DNA to function as a stable repository for information, as one of the few times chromosomes interact is during chromosomal crossover when they recombine.

Chromosomal crossover is when two DNA helices break, swap a section and then rejoin. Recombination allows chromosomes to exchange genetic information and produces new combinations of genes, which increases the efficiency of natural selection and can be important in the rapid evolution of new proteins. Genetic recombination can also be involved in DNA repair, particularly in the cell's response to double-strand breaks. The most common form of chromosomal crossover is homologous recombination, where the two chromosomes involved share very similar sequences.

Non-homologous recombination can be damaging to cells, as it can produce chromosomal translocations and genetic abnormalities. The recombination reaction is catalyzed by enzymes known as *recombinases,* such as RAD51. The first step in recombination is a double-stranded break either caused by an endonuclease or damage to the DNA.

A series of steps catalyzed in part by the recombinase then leads to joining of the two helices by at least one Holliday junction, in which a segment of a single strand in each helix is annealed to the complementary strand in the other helix. The Holliday junction is a tetrahedral junction structure that can be moved along the pair of chromosomes, swapping one strand for another. The recombination reaction is then halted by cleavage of the junction and re-ligation of the released DNA.

EVOLUTION OF DNA METABOLISM

DNA contains the genetic information that allows all modern living things to function, grow and reproduce. However, it is unclear how long in the 4-billion-year history of life DNA has performed this function, as it has been proposed that the earliest forms of life may have used RNA as their genetic material. RNA may have acted as the central part of early cell metabolism as it can both transmit genetic information and carry out catalysis as part of ribozymes.

This ancient RNA world where nucleic acid would have been used for both catalysis and genetics may have influenced the evolution of the current genetic code based on four nucleotide bases. This would occur since the number of unique bases in such an organism is a trade-off between a small number of bases increasing replication accuracy and a large number of bases increasing the catalytic efficiency of ribozymes. Unfortunately, there is no direct evidence of ancient genetic systems, as recovery of DNA from most fossils is impossible.

This is because DNA will survive in the environment for less than one million years and slowly degrades into short fragments in solution. Although claims for older DNA have been made, most notably a report of the isolation of a viable bacterium from a salt crystal 250-million years old, these claims are controversial and have been disputed.

Nucleases and Ligases

Nucleases are enzymes that cut DNA strands by catalyzing the hydrolysis of the phosphodiester bonds. Nucleases that hydrolyse nucleotides from the ends of DNA strands are called exonucleases, while endonucleases cut within strands.

The most frequently-used nucleases in molecular biology are the restriction endonucleases, which cut DNA at specific sequences. For instance, the EcoRV enzyme shown to the left recognizes the 6-base sequence 5' -GAT|ATC-3' and makes a cut at the vertical line.

In nature, these enzymes protect bacteria against phage infection by digesting the phage DNA when it enters the bacterial cell, acting as part of the restriction modification system. In technology, these sequence-specific nucleases are used in molecular cloning and DNA fingerprinting. Enzymes called DNA ligases can rejoin cut or broken DNA strands. Ligases are particularly important in lagging strand DNA replication, as they join together the short segments of DNA produced at the replication fork into a complete copy of the DNA template. They are also used in DNA repair and genetic recombination.

Topoisomerases and Helicases

Topoisomerases are enzymes with both nuclease and ligase activity. These proteins change the amount of supercoiling in DNA. Some of these enzyme work by cutting the DNA helix and allowing one section to rotate, thereby reducing its level of supercoiling; the enzyme then seals the DNA break. Other types of these enzymes are capable of cutting one DNA helix and then passing a second strand of DNA through this break, before rejoining the helix. Topoisomerases are required for many processes involving DNA, such as DNA replication and transcription. Helicases are proteins that are a type of molecular motor.

They use the chemical energy in nucleoside triphosphates, predominantly ATP, to break hydrogen bonds between bases and unwind the DNA double helix into single strands. These enzymes are essential for most processes where enzymes need to access the DNA bases.

Polymerases

Polymerases are enzymes that synthesize polynucleotide chains from nucleoside triphosphates. The sequence of their products are copies of existing polynucleotide chains - which are called *templates*. These enzymes function by adding nucleotides onto the 3' hydroxyl group of the previous nucleotide in a DNA strand. Consequently, all polymerases work in a 52 to 32 direction.

In the active site of these enzymes, the incoming nucleoside triphosphate base-pairs to the template: this allows polymerases to accurately synthesize the complementary strand of their template. Polymerases are classified according to the type of template that they use. In DNA replication, a DNA-dependent DNA polymerase makes a DNA copy of a DNA sequence.

Accuracy is vital in this process, so many of these polymerases have a proofreading activity. Here, the polymerase recognizes the occasional mistakes in the synthesis reaction by the lack of base pairing between the mismatched nucleotides.

If a mismatch is detected, a 3' to 5' exonuclease activity is activated and the incorrect base removed. In most organisms DNA polymerases function in a large complex called the replisome that contains multiple accessory subunits, such as the DNA clamp or helicases. RNA-dependent DNA polymerases are a specialized class of polymerases that copy the sequence of an RNA strand into DNA.

They include reverse transcriptase, which is a viral enzyme involved in the infection of cells by retroviruses, and telomerase, which is required for the replication of telomeres. Telomerase is an unusual polymerase because it contains its own RNA template as part of its structure.

Transcription is carried out by a DNA-dependent RNA polymerase that copies the sequence of a DNA strand into RNA. To begin transcribing a gene, the RNA polymerase binds to a sequence of DNA called a promoter and separates the DNA strands.

It then copies the gene sequence into a messenger RNA transcript until it reaches a region of DNA called the terminator, where it halts and detaches from the DNA. As with human DNA-dependent DNA polymerases, RNA polymerase II, the enzyme that transcribes most of the genes in the human genome, operates as part of a large protein complex with multiple regulatory and accessory subunits.

USES IN TECHNOLOGY

Genetic Engineering

Modern biology and biochemistry make intensive use of recombinant DNA technology. Recombinant DNA is a man-made DNA sequence that has been assembled from other DNA sequences. They can be transformed into organisms in the form of plasmids or in the appropriate format, by using a viral vector. The genetically modified organisms produced can be used to produce products such as recombinant proteins, used in medical research, or be grown in agriculture.

Forensics

Forensic scientists can use DNA in blood, semen, skin, saliva or hair at a crime scene to identify a perpetrator. This process is called genetic fingerprinting, or more accurately, DNA profiling. In DNA profiling, the lengths of variable sections of repetitive DNA, such as short tandem repeats and minisatellites, are compared between people. This method is usually an extremely reliable technique for identifying a criminal.

However, identification can be complicated if the scene is contaminated with DNA from several people. DNA profiling was developed in 1984 by British geneticist Sir Alec Jeffreys, and first used in forensic science to convict Colin Pitchfork in the 1988 Enderby murders case. People convicted of certain types of crimes may be required to provide a sample of DNA for a database.

This has helped investigators solve old cases where only a DNA sample was obtained from the scene. DNA profiling can also be used to identify victims of mass casualty incidents.

DNA Nanotechnology

The DNA structure at left will self-assemble into the structure visualized by atomic force microscopy at right. DNA nanotechnology is the field which seeks to design nanoscale structures using the molecular recognition properties of DNA molecules.

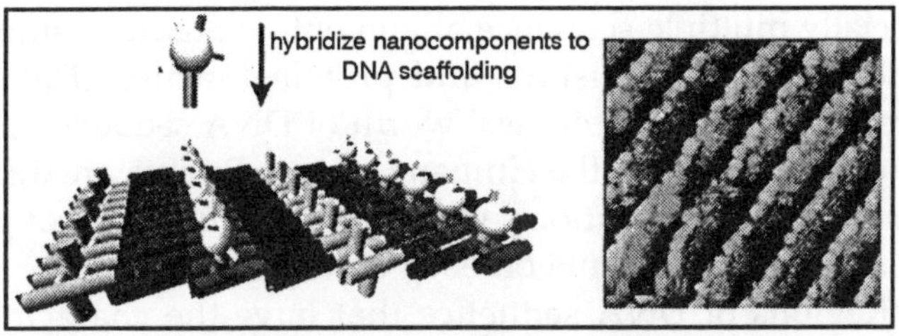

Fig. DNA Nanotechnology

DNA nanotechnology uses the unique molecular recognition properties of DNA and other nucleic acids to create self-assembling branched DNA complexes with useful properties. DNA is thus used as a structural material rather than as a carrier of biological information.

This has led to the creation of two-dimensional periodic lattices (both tile-based as well as using the "DNA origami" method) as well as three-dimensional structures in the shapes of polyhedra. Nanomechanical devices and algorithmic self-assembly have also been demonstrated, and these DNA structures have been used to template the arrangement of other molecules such as gold nanoparticles and streptavidin proteins.

Bioinformatics

Bioinformatics involves the manipulation, searching, and data mining of DNA sequence data. The development of techniques to store and search DNA sequences have led to widely-applied advances in computer science, especially string searching algorithms, machine learning and database theory.

String searching or matching algorithms, which find an occurrence of a sequence of letters inside a larger sequence of letters, were developed to search for specific sequences of nucleotides. In other applications such as text editors, even simple algorithms for this problem usually suffice, but DNA sequences cause these algorithms to exhibit near-worst-case behaviour due to their small number of distinct characters.

The related problem of sequence alignment aims to identify homologous sequences and locate the specific mutations that make them distinct. These techniques, especially multiple sequence alignment, are used in studying phylogenetic relationships and protein function. Data sets representing entire genomes' worth of DNA sequences, such as those produced by the Human Genome Project, are difficult to use without annotations, which label the locations of genes and regulatory elements on each chromosome.

Regions of DNA sequence that have the characteristic patterns associated with protein- or RNA-coding genes can be identified by gene finding algorithms, which allow researchers to predict the presence of particular gene products in an organism even before they have been isolated experimentally.

DNA FINGERPRINTING METHODS

DNA fingerprinting begins by extracting DNA from the cells in a sample of blood, saliva, semen, or other appropriate fluid or tissue.

STR Analysis

The most prevalent method of DNA fingerprinting used today is based on PCR and uses short tandem repeats (STR). This method uses highly polymorphic regions that have short repeated sequences of DNA (the most common is 4 bases repeated, but there are other lengths in use, including 3 and 5 bases).

Because different people have different numbers of repeat units, these regions of DNA can be used to discriminate between individuals. These STR loci (locations) are targeted with sequence-specific primers and are amplified using PCR. The DNA fragments that result are then separated and detected using electrophoresis.

There are two common methods of separation and detection, capillary electrophoresis (CE) and gel electrophoresis. The polymorphisms displayed at each STR region are by themselves very common, typically each polymorphism will be shared by around 5 - 20% of individuals. When looking at multiple loci, it is the unique combinations of these polymorphisms to an individual that makes this method discriminating as an identification tool.

The more STR regions that are tested in an individual the more discriminating the test becomes. From country to country different STR based DNA profiling systems are in use. In North America systems which amplify the CODIS 13 core loci are almost universal, while in the UK the SGM+ system, which is compatible with The National DNA Database in use. Whichever system is used, many of the STR regions under test are the same.

These DNA profiling systems are based around multiplex reactions, whereby many STR regions will be under test at the same time. Capillary electrophoresis works by electrokinetically (movement through the application of an electric field) injecting the DNA fragments into a thin glass tube (the capillary) filled with polymer. The DNA is pulled through the tube by the application of an electric field, separating the fragments such that the smaller fragments travel faster through the capillary.

The fragments are then detected using fluorescent dyes that were attached to the primers used in PCR. This allows multiple fragments to be amplified and run simultaneously, something known as multiplexing. Sizes are assigned using labeled DNA size standards that are added to each sample, and the number of repeats are determined by comparing the size to an allelic ladder, a sample that contains all of the common possible repeat sizes. Although this method is expensive, larger capacity machines with higher throughput are being used to lower the cost/sample and reduce backlogs that exist in many government crime facilities. Gel electrophoresis acts using similar principles as CE, but instead of using a capillary, a large polyacrylamide gel is used to separate the DNA fragments.

An electric field is applied, as in CE, but instead of running all of the samples by a detector, the smallest fragments are run close to the bottom of the gel and the entire gel is scanned into a computer. This produces an image showing all of the bands corresponding to different repeat sizes and the allelic ladder. This approach does not require the use of size standards, since the allelic ladder is run alongside the samples and serves this purpose. Visualization can either be through the use of fluorescently tagged dyes in the primers or by silver staining the gel prior to scanning.

Although it is cost effective and can be rather high throughput, silver staining kits for STRs are being discontinued. In addition, many labs are phasing out gels in favour of CE as the cost of machines becomes more manageable.

The true power of STR analysis is in its statistical power of discrimination. In the U.S.A., there are 13 core loci (DNA locations) that are currently used for discrimination in CODIS. Because these loci are independently assorted (having a certain number of repeats at one locus doesn't change the likelihood of having any number of repeats at any other locus), the product rule for probabilities can be applied.

This means that if someone has the DNA type of ABC, where the three loci were independent, we can say that the probability of having that DNA type is the probability of having type A times the probability of having type B times the probability of having type C.

This has resulted in the ability to generate match probabilities of 1 in a quintillion (1 with 18 zeros after it) or more. At least, that is the theory.

Within the relevant small subpopulation, the probabilities of the different types could be very very different from what they are in the whole population. Instead of multiplying 13 times a smallish probability together, and coming up with one in a quintillion, probably one should multiply together 13 times a quite large but quite unknown probability - resulting is something like 1 in 10, or 1 in a thousand? No-one knows.

There is presently much concern among scientists that the lack of understanding of DNA evidence among police, lawyers, and judges, has already caused serious miscarriages of justice in many countries.

DNA evidence is useful for showing that someone could not be guilty (when you have a mismatch), but its value in proving that someone is guilty, is basically dubious, and unknown.

RFLP Analysis

One way to fingerprint DNA is by doing a Southern blot. This has several steps. First, the DNA being analysed must be separated from other material. Next, it must be cut into a few different-sized pieces using restriction enzymes, proteins that can cut double-stranded DNA without damaging the bases. The pieces are sorted by size through gel electrophoresis. The pieces are poured into gel with a positive charge at the bottom.

DNA has a natural slightly negative charge so it will be attracted to the bottom. The smaller pieces can move more quickly through the gel, therefore they will be further toward the bottom than the larger pieces. This will separate the pieces by size, with the larger ones higher up and the smaller ones further down. Next, alkaline solution or heat is applied to the gel so that the DNA denatures and separates into single strands. Nitrocellulose paper is pressed evenly against the gel and then baked so the DNA is permanently attached to it.

The DNA is now ready to be analysed using a radioactive probe in a hybridization reaction. To make a radioactive probe, DNA polymerase is needed. The DNA that is going to be made radioactive should be put in a tube. Horizontal breaks should be made along the strand, while at the same time nucleotides should be added. The base C, or cytosine, should be radioactive. Next, the polymerase should be added to the tube. It will be attracted to the breaks and try to fix them. As the DNA polymerase fixes the DNA, it will break the existing bonds so that the existing nucleotides can be replaced by the new nucleotides in the tube.

Whenever the lower strand has a G base, or guanine, the C put in will be radioactive. By repairing the strand of DNA, the polymerase is also making it radioactive. The DNA is heated so that the two strands split. Single-stranded pieces that might or might not be radioactive are made. The radioactive pieces are now probes ready for use. Now the radioactive probe can be used to create a hybridization reaction. Hybridization is when two genetic sequences bind together because of the hydrogen bonds that are in between the base pairs.

There are two of these bonds between A, or adenine, and T, or thymine, and three between C and G. To make hybridization works, the DNA has to be denatured so it is single-stranded; like the Southern Blot that was made on the nitrocellulose paper.

The denatured DNA and the radioactive probe should be put into a plastic bag with saline liquid, and then shaken. The probe will bond to the denatured DNA wherever it finds a fit. The probe and the DNA do not have to fit together precisely. The two will have sequences that can stick together even if the fit is poor, however there will be fewer hydrogen bonds.

Probes that have low homology, or similarity, can bind to the DNA better if the temperature is varied or the amount of salt in the mixture is changed. Even if the fit is poor, the probe and the DNA are now hybridized. A way to make use of the whole process described is by using it to determine a person's VNTRs. VNTRs, or Variable Number Tandem Repeats, are repeated sequences of base pairs in someone's genetic information. Every DNA strand contains exons, or sections that have genetic information, and introns, which have no discernible use other than containing VNTRs, or repeating sequences of base pairs.

Every single human being has a few of these repeating sequences. To find out if somebody has a specific VNTR, a Southern Blot must be made, and then probed in a hybridization reaction, by a radioactive version of said VNTR. This process ends up making a pattern called a DNA fingerprint. Every person has VNTRs they have inherited genetically from one or both parents. It is impossible for somebody to have one that neither of their parents did. VNTR patterns are unique for each person, and they will be more exact if more VNTR probes are used.

PCR Analysis

With the invention of the polymerase chain reaction (PCR), DNA fingerprinting took huge strides forward in both discriminating power and ability to recover information from very small starting samples. PCR involves the amplification of specific regions of DNA using a cycling of temperature and a thermostable polymerase enzyme along with flourescently labelled sequence specific primers of DNA. Commercial kits that used single nucleotide polymorphisms (SNPs) for discrimination became available.

These kits use PCR to amplify specific regions with known variations and hybridize them to probes anchored on cards, which results in a coloured spot corresponding to the particular sequence variation. One of the primary complaints against RFLP was that it was slow and required large quantities of DNA to be used. This led to the development of PCR-based methods which required smaller amounts of DNA that could also be more degraded than those used in RFLP analysis.

Systems such as the HLA-DQ alpha reverse dot blot strips grew to be very popular due to their ease of use and the speed with which a result could be obtained, however they were not as discriminating as RFLP. It was also difficult to determine a DNA profile for mixed samples, such as a vaginal swab from a sexual assault victim.

AmpFLP

Another technique, AmpFLP, or amplified fragment length polymorphism was also put into practice during the early 1990s. This technique was also faster than RFLP analysis and used PCR to amplify DNA samples. It relied on variable number tandem repeat (VNTR) polymorphisms to distinguish various alleles, which were separated on a polyacrylamide gel using an allelic ladder (as opposed to a molecular weight ladder). Bands could be visualized by silver staining the gel.

One popular locus for fingerprinting was the D1S80 locus. As with all PCR based methods, highly degraded DNA or very small amounts of DNA may cause allelic dropout (causing a mistake in thinking a heterozygote is a homozygote) or other stochastic effects. In addition, because the analysis is done on a gel, very high number repeats may bunch together at the top of the gel, making it difficult to resolve.

AmpFLP analysis can be highly automated, and allows for easy creation of phylogenetic trees based on comparing individual samples of DNA. Due to its relatively low cost and ease of set-up and operation, AmpFLP remains popular in lower income countries.

Y-chromosome Analysis

Recent innovations have included the creation of primers targeting polymorphic regions on the Y-chromosome (Y-STR), which allows resolution of multiple male profiles, or cases in which a differential extraction is not possible. Y-chromosomes are paternally inherited, so Y-STR analysis can help in the identification of paternally related males. Y-STR analysis was performed in the Sally Hemings controversy to determine if Thomas Jefferson had sired a son with one of his slaves.

Mitochondrial Analysis

For highly degraded samples, it is sometimes impossible to get a complete profile of the 13 CODIS STRs. In these situations, mitochondrial DNA (mtDNA) is sometimes typed due to there being many copies of mtDNA in a cell, while there may only be 1-2 copies of the nuclear DNA. Forensic scientists amplify the HV1 and HV2 regions of the mtDNA, then sequence each region and compare single nucleotide differences to a reference.

Because mtDNA is maternally inherited, directly linked maternal relatives can be used as match references, such as one's maternal grandmother's sister's son.

A difference of two or more nucleotides is generally considered to be an exclusion. Heteroplasmy and poly-C differences may throw off straight sequence comparisons, so some expertise on the part of the analyst is required. mtDNA is useful in determining unclear identities, such as those of missing persons when a maternally linked relative can be found. mtDNA testing was used in determining that Anna Anderson was not the Russian princess she had claimed to be, Anastasia Romanov. mtDNA can be obtained from such material as hair shafts and old bones/teeth.

Evaluating DNA Evidence

In the early days of the use of genetic fingerprinting as criminal evidence, juries were often swayed by spurious statistical arguments by defence lawyers along these lines: given a match that had a 1 in 5 million probability of occurring by chance, the lawyer would argue that this meant that in a country of say 60 million people there were 12 people who would also match the profile. This was then translated to a 1 in 12 chance of the suspect being the guilty one. This argument is not sound unless the suspect was drawn at random from the population of the country.

In fact, a jury should consider how likely it is that an individual matching the genetic profile would also have been a suspect in the case for other reasons. Another spurious statistical argument is based on the false assumption that a 1 in 5 million probability of a match automatically translates into a 1 in 5 million probability of guilt and is known as the prosecutor's fallacy. When using RFLP, the theoretical risk of a coincidental match is 1 in 100 billion (100,000,000,000). However, the rate of laboratory error is almost certainly higher than this, and often actual laboratory procedures do not reflect the theory under which the coincidence probabilities were computed.

The coincidence probabilities may be calculated based on the probabilities that markers in two samples have bands in *precisely* the same location, but a laboratory worker may conclude that similar—but not precisely identical—band patterns result from identical genetic samples with some imperfection in the agarose gel. However, in this case, the laboratory worker increases the coincidence risk by expanding the criteria for declaring a match. Recent studies have quoted relatively high error rates which may be cause for concern. In the early days of genetic fingerprinting, the necessary population data to accurately compute a match probability was sometimes unavailable.

Between 1992 and 1996, arbitrary low ceilings were controversially put on match probabilities used in RFLP analysis rather than the higher theoretically computed ones.

Today, RFLP has become widely disused due to the advent of more discriminating, sensitive and easier technologies. STRs do not suffer from such subjectivity and provide similar power of discrimination.

It should be noted that figures of this magnitude are not considered to be statistically supportable by scientists, for unrelated individuals with full matching DNA profiles a match probability of 1 in a billion (one thousand million) is considered statistically supportable. However, with any DNA technique, the cautious juror should not convict on genetic fingerprint evidence alone if other factors raise doubt.

Contamination with other evidence (secondary transfer) is a key source of incorrect DNA profiles and raising doubts as to whether a sample has been adulterated is a favourite defence technique. More rarely, Chimerism is one such instance where the lack of a genetic match may unfairly exclude a suspect.

Chapter 9

Medical Ultrasonography

In medical imaging ultrasound plays an important role, which has numerous applications in imaging abdominal, breast, musculoskeletal, obstetric, pelvic, prostate, thyroid and vascular tissues. New developments such as three- and four-dimensional ultrasound imaging systems1 are especially useful in knowing the foetal health-status.

These images are critical in case of any biopsy needed for the foetus. With these new techniques, parents can see the three-dimensional foetal images that reveal the facial expressions. The sad part of this ultrasound imaging is that it is being used for sex determination and discrimination in most parts of India.

The recent census shows that the sex ratio (females: males) among children in the 0–6 age group is 927/1000 which is low compared to the 1991 census (945/ 1000). This ratio is one of the world's lowest; the statistical norm is 1050 females for every 1000 males. The number of ultrasound centres throughout India is growing rapidly. Many officials agree that the Pre-Natal Diagnostic Techniques (PNDT) (Regulation and Prevention of Misuse) act passed in 1994, has not been effective in stopping the girl child extermination. Most of the ultrasound centres in India are unregistered, even though PNDT requires all the ultrasound centres to be registered. The alarming decrease in this sex-ratio needs serious thinking by officials and the medical community.

Ethical issues of ultrasound imaging and there is an argument that the foetal images seen by the mother encourage her to stay healthy during pregnancy and may establish a stronger bond which improves the health of the baby and the mother.

Hence people are encouraged to have these scans. We have another problem in ultrasound imaging when these examinations are carried out by untrained staff. Such examinations are not good as the foetus may be exposed to more ultrasonic waves than necessary.

Ultrasound energy is potentially harmful, even though till now there is no evidence in establishing the biological effect of this technology; harmful effects could be discovered in future. There must be awareness campaigns on these issues of sex determination and ultrasound imaging by NGOs and government officials. The question of whether ultrasound is a boon or bane reverts to the same dilemma as to whether technology is making us better or bitter.

BIOCHEMICAL ENGINEERS

Biochemical engineers, a fraternity into which are expressing increasing anguish that they are becoming professionally marginalized. Like all aggrieved groups, they blame other professions, mainly biologists and chemical engineers, for their alienation. There may be some truth in these accusations. Biologists in academic and research organizations are reluctant to acknowledge the contributions of biochemical engineers to biotechnology.

Paradoxically, they also complain that biochemical engineers do not easily integrate with biologists. This self-contradiction has resulted in most biotechnology faculties in India being devoid of biochemical engineers. While accepting biochemical engineers nominally, chemical engineering bodies relegate them to fringe activities. Thus, biochemical engineering gets only minor roles in chemical engineering curricula, conferences and symposia. However, biochemical engineers may have to share some of the blame for their own predicament.

Since theirs is a relatively young profession that straddles biology and chemical engineering, they have the difficult task of blending the two older disciplines and evolving new concepts. This will take both time and a change in their mindsets.

To avoid the danger of their profession becoming just a hybrid of biology and chemical engineering, biochemical engineers need to absorb from both disciplines and venture into niche areas such as metabolic and immunological engineering and lab-on-chip technologies, where their special skills will propel them toward the recognition and respect they deserve. Presently the research group is focused upon design and development of a low power, low cost, high performance embedded ultrasound imaging system.

The system has three parts; an analog front-end, a digital signal processing hardware and a software application programme. The signal processing part is implemented using Xilinx FPGAs, due to the need for both real time operation and further reconfigurability, if required. Extensive parallelism and pipelining have been exploited while mapping the algorithms into VLSI. The system board is fully integrable with a standard PC.

The user can see both real time video and individual frames. Also, the hardware has a unit which performs compression of image frames and storing them in the hard disk. The control and data handling are taken care of by the software. The proposed system has got a 64 element phased array transducer at the very front end.

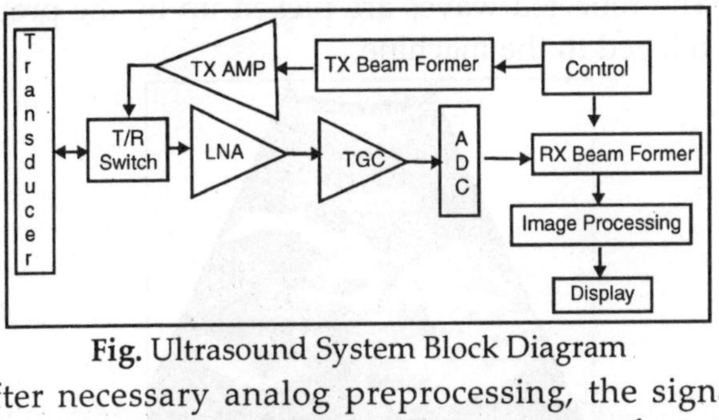

Fig. Ultrasound System Block Diagram

After necessary analog preprocessing, the signals are digitized and sent to the FPGAs. The system performs digital beamforming by properly delaying and weighting signals coming from all 64 channels, and then adding them to get a coherent output. The amplitude and phase modulation parameters are then extracted to obtain the intensity value. Since the beamformed data is in polar format, a scan conversion is required before displaying the pixels on a TV raster display.

Also, ultrasound images are greatly corrupted with noise, hence some image filtering operations are also done. The image frames are transformed and coded before storing them in the hard disk of the PC.

ULTRASONOGRAPHY

Ultrasonography or Ultrasound is a medical imaging technique that uses high frequency sound waves and their echoes. The technique is similar to the echolocation used by bats, whales and dolphins, as well as SONAR used by submarines. In ultrasound, the following events happen:

- The ultrasound machine transmits high-frequency (1 to 5 megahertz) sound pulses into your body using a probe.
- The sound waves travel into your body and hit a boundary between tissues (e.g. between fluid and soft tissue, soft tissue and bone).
- Some of the sound waves get reflected back to the probe, while some travel on further until they reach another boundary and get reflected.
- The reflected waves are picked up by the probe and relayed to the machine.

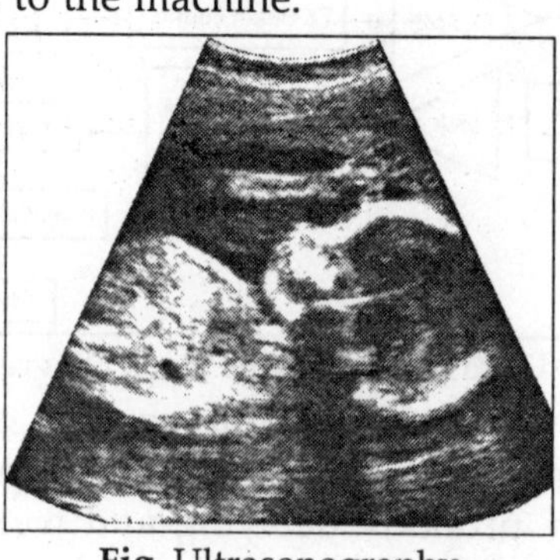

Fig. Ultrasonography

- The machine calculates the distance from the probe to the tissue or organ (boundaries) using the speed of sound in tissue (5,005 ft/s or1,540 m/s) and the time of the each echo's return (usually on the order of millionths of a second).
- The machine displays the distances and intensities of the echoes on the screen, forming a two dimensional image.

How Ultrasonography Works

There are many situations in which ultrasound is performed. Perhaps you are pregnant, and your obstetrician wants you to have an ultrasound to check on the developing baby or determine the due date.

Maybe you are having problems with blood circulation in a limb or your heart, and your doctor has requested a Doppler ultrasound to look at the blood flow. Ultrasound has been a popular medical imaging technique for many years.

Types of Ultrasonography

The ultrasound that we have described so far presents a two dimensional image, or "slice," of a three dimensional object (fetus, organ). Two other types of ultrasound are currently in use, 3D ultrasound imaging and Doppler ultrasound.

3D Ultrasonography Imaging

In the past two years, ultrasound machines capable of three-dimensional imaging have been developed. In these machines, several two-dimensional images are acquired by moving the probes across the body surface or rotating inserted probes.

The two-dimensional scans are then combined by specialized computer software to form 3D images.

- Early detection of cancerous and benign tumors
- Examining the prostate gland for early detection of tumors

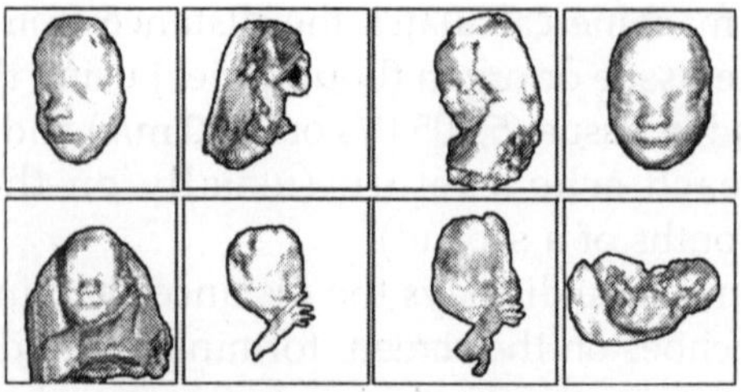

Fig. 3D Imaging Allows you to Get a Better Look at the Organ Being Examined and is Best Used for:

- Looking for masses in the colon and rectum
- Detecting breast lesions for possible biopsies
- Visualizing a fetus to assess its development, especially for observing abnormal development of the face and limbs
- Visualizing blood flow in various organs or a fetus

Doppler Ultrasonography

Doppler ultrasound is based upon the Doppler Effect. When the object reflecting the ultrasound waves is moving, it changes the frequency of the echoes, creating a higher frequency if it is moving toward the probe and a lower frequency if it is moving away from the probe. How much the frequency is changed depends upon how fast the object is moving.

Doppler ultrasound measures the change in frequency of the echoes to calculate how fast an object is moving. Doppler ultrasound has been used mostly to measure the rate of blood flow through the heart and major arteries.

USES OF ULTRASONOGRAPHY

Ultrasound has been used in a variety of clinical settings, including obstetrics and gynecology, cardiology and cancer detection.

The main advantage of ultrasound is that certain structures can be observed without using radiation. Ultrasound can also be done much faster than X-rays or other radiographic techniques. Here is a short list of some uses for ultrasound:

Obstetrics and Gynecology

- Measuring the size of the fetus to determine the due date
- Determining the position of the fetus to see if it is in the normal head down position or breech
- Checking the position of the placenta to see if it is improperly developing over the opening to the uterus (cervix)
- Seeing the number of fetuses in the uterus
- Checking the fetus's growth rate by making many measurements over time
- Detecting ectopic pregnancy, the life-threatening situation in which the baby is implanted in the mother's Fallopian tubes instead of in the uterus
- Determining whether there is an appropriate amount of amniotic fluid cushioning the baby
- Monitoring the baby during specialized procedures - ultrasound has been helpful in seeing and avoiding the baby during amniocentesis (sampling of the amniotic fluid with a needle for genetic testing). Years ago, doctors use to perform this procedure blindly; however, with accompanying use of ultrasound, the risks of this procedure have dropped dramatically.
- Seeing tumors of the ovary and breast

Cardiology

- Seeing the inside of the heart to identify abnormal structures or functions
- Measuring blood flow through the heart and major blood vessels

Urology

- Measuring blood flow through the kidney
- Seeing kidney stones
- Detecting prostate cancer early
- In addition to these areas, there is a growing use for ultrasound as a rapid imaging tool for diagnosis in emergency rooms.

MEDICAL ULTRASONOGRAPHY

Medical sonography (ultrasonography) is an ultrasound-based diagnostic imaging technique used to visualize muscles and internal organs, their size, structures and possible pathologies or lesions. Obstetric sonography is commonly used during pregnancy and is widely recognized by the public. There are a plethora of diagnostic and therapeutic applications practiced in medicine. In physics the term "ultrasound" applies to all acoustic energy with a frequency above human hearing (20,000 hertz or 20 kilohertz).

Typical diagnostic sonographic scanners operate in the frequency range of 2 to 18 megahertz, hundreds of times greater than the limit of human hearing. The choice of frequency is a trade-off between spatial resolution of the image and imaging depth: lower frequencies produce less resolution but image deeper into the body.

Diagnostic Applications

Sonography (ultrasonography) is widely used in medicine. It is possible to perform diagnosis or therapeutic procedures with the guidance of sonography (for instance biopsies or drainage of fluid collections). Sonographers are medical professionals who perform scans for diagnostic purposes. Sonographers typically use a hand-held probe (called a transducer) that is placed directly on and moved over the patient.

A water-based gel is used to couple the ultrasound between the transducer and patient. Sonography is effective for imaging soft tissues of the body. Superficial structures such as muscles, tendons, testes, breast and the neonatal brain are imaged at a higher frequency (7-18 MHz), which provides better axial and lateral resolution.

Deeper structures such as liver and kidney are imaged at a lower frequency 1-6 MHz with lower axial and lateral resolution but greater penetration.

Medical sonography is used in:

- Cardiology
- Endocrinology

- Gastroenterology
- Gynaecology
- Obstetrics
- Ophthalmology
- Urology
- Musculoskeletal, tendons, muscles, and nerves
- Vascular, arteries and veins
- Intravascular ultrasound
- Intervenional; biopsy, emptying fluids, intrauterine transfusion
- Contrast-enhanced ultrasound

A general-purpose sonographic machine may be able to be used for most imaging purposes. Usually specialty applications may be served only by use of a specialty transducer. The dynamic nature of many studies generally requires specialized features in a sonographic machine for it to be effective; such as endovaginal, endorectal, or transesophageal transducers.

Obstetrical ultrasound is commonly used during pregnancy to check on the development of the fetus. In a pelvic sonogram, organs of the pelvic region are imaged. This includes the uterus and ovaries or urinary bladder. Men are sometimes given a pelvic sonogram to check on the health of their bladder and prostate.

There are two methods of performing a pelvic sonography - externally or internally. The internal pelvic sonogram is performed either transvaginally (in a woman) or transrectally (in a man).

Sonographic imaging of the pelvic floor can produce important diagnostic information regarding the precise relationship of abnormal structures with other pelvic organs and it represents a useful hint to treat patients with symptoms related to pelvic prolapse, double incontinence and obstructed defecation.

In abdominal sonography, the solid organs of the abdomen such as the pancreas, aorta, inferior vena cava, liver, gall bladder, bile ducts, kidneys, and spleen are imaged. Sound waves are blocked by gas in the bowel, therefore there are limited diagnostic capabilities in this area. The appendix can sometimes be seen when inflamed eg: appendicitis.

Therapeutic Applications

Therapeutic applications use ultrasound to bring heat or agitation into the body. Therefore much higher energies are used than in diagnostic ultrasound. In many cases the range of frequencies used are also very different.

- Ultrasound may be used to clean teeth in dental hygiene.
- Ultrasound sources may be used to generate regional heating in biological tissue, e.g. in occupational therapy, physical therapy and cancer treatment.
- Focused ultrasound may be used to generate highly localized heating to treat cysts and tumors (benign or malignant), This is known as Focused Ultrasound Surgery (FUS) or High Intensity Focused Ultrasound (HIFU). These procedures generally use lower frequencies than medical diagnostic ultrasound (from 250 kHz to 2000 kHz), but significantly higher energies. HIFU treatment is often guided by MRI.
- Focused ultrasound may be used to break up kidney stones by lithotripsy.
- Ultrasound may be used for cataract treatment by phacoemulsification.
- Additional physiological effects of low-intensity ultrasound have recently been discovered, e.g. its ability to stimulate bone-growth and its potential to disrupt the blood-brain barrier for drug delivery.

From Sound to Image

The creation of an image from sound is done in three steps - producing a sound wave, receiving echoes, and interpreting those echoes.

Producing a Sound Wave

A sound wave is typically produced by a piezoelectric transducer encased in a probe. Strong, short electrical pulses from the ultrasound machine make the transducer ring at the desired frequency. The frequencies can be anywhere between 2 and 15 MHz. The sound is focused either by the shape of the transducer, a lens in front of the transducer, or a complex set of control pulses from the ultrasound scanner machine. This focusing produces an arc-shaped sound wave from the face of the transducer.

The wave travels into the body and comes into focus at a desired depth. Older technology transducers focus their beam with physical lenses. Newer technology transducers use phased array techniques to enable the sonographic machine to change the direction and depth of focus. Almost all piezoelectric transducers are made of ceramic. Materials on the face of the transducer enable the sound to be transmitted efficiently into the body (usually seeming to be a rubbery coating, a form of impedance matching).

In addition, a water-based gel is placed between the patient's skin and the probe. The sound wave is partially reflected from the layers between different tissues. Specifically, sound is reflected anywhere there are density changes in the body: e.g. blood cells in blood plasma, small structures in organs, etc. Some of the reflections return to the transducer.

Receiving the Echoes

The return of the sound wave to the transducer results in the same process that it took to send the sound wave, except in reverse. The return sound wave vibrates the transducer, the transducer turns the vibrations into electrical pulses that travel to the ultrasonic scanner where they are processed and transformed into a digital image.

Forming the Image

The sonographic scanner must determine three things from each received echo:

- How long it took the echo to be received from when the sound was transmitted.
- From this the focal length for the phased array is deduced, enabling a sharp image of that echo at that

depth (this is not possible while producing a sound wave).

- How strong the echo was. It could be noted that sound wave is no click, but a pulse with a specific carrier frequency. Moving objects change this frequency on reflection, so that it is only a matter of electronics to have simultaneous Doppler sonography.

Once the ultrasonic scanner determines these three things, it can locate which pixel in the image to light up and to what intensity and at what hue if frequency is processed.

Transforming the received signal into a digital image may be explained by using a blank spreadsheet as an analogy. We imagine our transducer is a long, flat transducer at the top of the sheet. We send pulses down the 'columns' of our spreadsheet (A, B, C, etc.). We listen at each column for any return echoes.

When we hear an echo, we note how long it took for the echo to return. The longer the wait, the deeper the row (1,2,3, etc.). The strength of the echo determines the brightness setting for that cell (white for a strong echo, black for a weak echo, and varying shades of grey for everything in between.) When all the echoes are recorded on the sheet, we have a greyscale image.

Sound in the Body

Ultrasonography (sonography) uses a probe containing one or more acoustic transducers to send pulses of sound into a material. Whenever a sound wave encounters a material with a different density (acoustical impedance), part of the sound wave is reflected back to the probe and is detected as an echo.

The time it takes for the echo to travel back to the probe is measured and used to calculate the depth of the tissue interface causing the echo. The greater the difference between acoustic impedances, the larger the echo is.

If the pulse hits gases or solids, the density difference is so great that most of the acoustic energy is reflected and it becomes impossible to see deeper. The frequencies used for medical imaging are generally in the range of 1 to 18 MHz.

Higher frequencies have a correspondingly smaller wavelength, and can be used to make sonograms with smaller details. However, the attenuation of the sound wave is increased at higher frequencies, so in order to have better penetration of deeper tissues, a lower frequency (3-5 MHz) is used.

Seeing deep into the body with sonography is very difficult. Some acoustic energy is lost every time an echo is formed, but most of it (approximately)is lost from acoustic absorption. The speed of sound is different in different materials, and is dependent on the acoustical impedance of the material.

However, the sonographic instrument assumes that the acoustic velocity is constant at 1540 m/s. An effect of this assumption is that in a real body with non-uniform tissues, the beam become somewhat de-focused and image resolution is reduced.

To generate a 2D-image, the ultrasonic beam is swept. A transducer may be swept mechanically by rotating or swinging. Or a 1D phased array transducer may be use to sweep the beam electronically.

The received data is processed and used to construct the image. The image is then a 2D representation of the slice into the body. 3D images can be generated by acquiring a series of adjacent 2D images. Commonly a specialised probe that mechanically scans a conventional 2D-image transducer is used.

However, since the mechanical scanning is slow, it is difficult to make 3D images of moving tissues. Recently, 2D phased array transducers that can sweep the beam in 3D have been developed. These can image faster and can even be used to make live 3D images of a beating heart. Doppler ultrasonography is used to study blood flow and muscle motion.

The different detected speeds are represented in colour for ease of interpretation, leaky heart valves: the leak shows up as a flash of unique colour. Colours may alternatively be used to represent the amplitudes of the received echoes.

Modes of Sonography

Four different modes of ultrasound are used in medical imaging. These are:

- *A-mode*: A-mode is the simplest type of ultrasound. A single transducer scans a line through the body with the echoes plotted on screen as a function of depth. Therapeutic ultrasound aimed at a specific tumor or calculus is also A-mode, to allow for pinpoint accurate focus of the destructive wave energy.
- *B-mode*: In B-mode ultrasound, a linear array of transducers simultaneously scans a plane through the body that can be viewed as a two-dimensional image on screen.
- *M-mode*: M stands for motion. In m-mode a rapid sequence of B-mode scans whose images follow each other in sequence on screen enables doctors to see and measure range of motion, as the organ boundaries that produce reflections move relative to the probe.
- *Doppler mode*: This mode makes use of the Doppler effect.

Doppler Sonography

Sonography can be enhanced with Doppler measurements, which employ the Doppler effect to assess whether structures (usually blood) are moving towards or away from the probe, and its relative velocity. By calculating the frequency shift of a particular sample volume, a jet of blood flow over a heart valve, its speed and direction can be determined and visualised. This is particularly useful in cardiovascular studies (sonography of the vasculature system and heart) and essential in many areas such as determining reverse blood flow in the liver vasculature in portal hypertension.

The Doppler information is displayed graphically using spectral Doppler, or as an image using colour Doppler (directional Doppler) or power Doppler (non directional Doppler). This Doppler shift falls in the audible range and is often presented audibly using stereo speakers: this produces a very distinctive, although synthetic, pulsing sound. Strictly speaking, most modern sonographic machines do not use the Doppler effect to measure velocity, as they rely on pulsed wave Doppler (PW). Pulsed wave machines transmit pulses of ultrasound, and then switch to receive mode.

As such, the reflected pulse that they receive is not subject to a frequency shift, as the insonation is not continuous. However, by making several measurements, the phase change in subsequent measurements can be used to obtain the frequency shift (since frequency is the rate of change of phase).

To obtain the phase shift between the received and transmitted signals, one of two algorithms is typically used: the Kasai algorithm or cross-correlation. Older machines, that use continuous wave (CW) Doppler, exhibit the Doppler effect.

To do this, they must have separate transmission and reception transducers. The major drawback of CW machines, is that no distance information can be obtained (this is the major advantage of PW systems - the time between the transmitted and received pulses can be converted into a distance with knowledge of the speed of sound). In the sonographic community (although not in the signal processing community), the terminology "Doppler" ultrasound, has been accepted to apply to both PW and CW Doppler systems despite the different mechanisms by which the velocity is measured.

Microbubbles

The use of microbubble contrast media in medical sonography to improve ultrasound signal backscatter is known as contrast-enhanced ultrasound. This technique is currently used in echocardiography, and may have future applications in molecular imaging and drug delivery.

Strengths of Sonography

- It images muscle and soft tissue very well and is particularly useful for delineating the interfaces between solid and fluid-filled spaces.

- It renders "live" images, where the operator can dynamically select the most useful section for diagnosing and documenting changes, often enabling rapid diagnoses.
- It shows the structure of organs.
- It has no known long-term side effects and rarely causes any discomfort to the patient.
- Equipment is widely available and comparatively flexible.
- Small, easily carried scanners are available; examinations can be performed at the bedside.
- Relatively inexpensive compared to other modes of investigation (e.g. computed X-ray tomography, DEXA or magnetic resonance imaging).

Weaknesses of Ultrasonic Imaging

- Sonographic devices have trouble penetrating bone. Sonography of the adult brain is very limited.
- Sonography performs very poorly when there is a gas between the transducer and the organ of interest, due to the extreme differences in acoustic impedance. Overlying gas in the gastrointestinal tract often makes ultrasound scanning of the pancreas difficult, and lung imaging is not possible (apart from demarcating pleural effusions).
- Even in the absence of bone or air, the depth penetration of ultrasound is limited, making it difficult to image structures deep in the body, especially in obese patients.
- The method is operator-dependent. A high level of skill and experience is needed to acquire good-quality images and make accurate diagnoses.
- There is no scout image as there is with CT and MR. Once an image has been acquired there is no exact way to tell which part of the body was imaged.

Risks and Side-effects

Ultrasonography is generally considered a "safe" imaging

modality. However slight detrimental effects have been occasionally observed.

Diagnostic ultrasound studies of the fetus are generally considered to be safe during pregnancy. This diagnostic procedure should be performed only when there is a valid medical indication, and the lowest possible ultrasonic exposure setting should be used to gain the necessary diagnostic information under the "as low as reasonably achievable" or ALARA principle.

The promotion, selling, or leasing of ultrasound equipment for making "keepsake fetal videos" is considered by the US Food and Drug Administration to be an unapproved use of a medical device. The gap between intended medical use and the increasingly common patient desired "keepsake" format has led to the development of a service called PhotOBaby available only in medical offices.

Studies on the Safety of Ultrasound

- A study at the Yale Medical School found a correlation between prolonged and frequent use of ultrasound and abnormal neuronal migration in mice.
- A study published in 2001 by a team working at the Karolinska Institute in Stockholm found a correlation between the number of scans received by male fetuses and subsequent left-handedness.
- A meta-analysis of several ultrasonography studies found no statistically significant harmful effects from ultrasonography, but mentioned that there was a lack of data on long-term substantive outcomes such as neurodevelopment.

OBSTETRIC ULTRASOUND SCANS

Obstetric Ultrasound is the use of ultrasound scans in pregnancy. Since its introduction in the late 1950's ultrasonography has become a very useful diagnostic tool in Obstetrics.

Currently used equipments are known as real-time scanners, with which a continous picture of the moving fetus can be depicted on a monitor screen. Very high frequency sound waves of between 3.5 to 7.0 megahertz (i.e. 3.5 to 7 million cycles per second) are generally used for this purpose.

They are emitted from a transducer which is placed in contact with the maternal abdomen, and is moved to "look at" (likened to a light shined from a torch) any particular content of the uterus. Repetitive arrays of ultrasound beams scan the fetus in thin slices and are reflected back onto the same transducer.

The information obtained from different reflections are recomposed back into a picture on the monitor screen (a sonogram, or ultrasonogram).

Movements such as fetal heart beat and malformations in the feus can be assessed and measurements can be made accurately on the images displayed on the screen. Such measurements form the cornerstone in the assessment of gestational age, size and growth in the fetus.

A full bladder is often required for the procedure when abdominal scanning is done in early pregnency. There may be some discomfort from pressure on the full bladder. The conducting gel is non-staining but may feel slightly cold and wet.

There is no sensation at all from the ultrasound waves. Ultrasound scan is currently considered to be a safe, non-invasive, accurate and cost-effective investigation in the fetus. It has progressively become an indispensible obstetric tool and plays an important role in the care of every pregnant woman.

USE OF ULTRASONOGRAPHY

DIAGNOSIS AND CONFIRMATION OF EARLY PREGNANCY

The gestational sac can be visualized as early as four and a half weeks of gestation and the yolk sac at about five weeks. The embryo can be observed and measured by about five and a half weeks. Ultrasound can also very importantly confirm the site of the pregnancy is within the cavity of the uterus.

VAGINAL BLEEDING IN EARLY PREGNANCY

The viability of the fetus can be documented in the presence of vaginal bleeding in early pregnancy. A visible heartbeat could be seen and detectable by pulsed doppler ultrasound by about 6 weeks and is usually clearly depictable by 7 weeks. If this is observed, the probability of a continued pregnancy is better than 95 percent. Missed abortions and blighted ovum will usually give typical pictures of a deformed gestational sac and absence of fetal poles or heart beat.

Fetal heart rate tends to vary with gestational age in the very early parts of pregnancy. Normal heart rate at 6 weeks is around 90-110 beats per minute (bpm) and at 9 weeks is 140-170 bpm.

At 5-8 weeks a bradycardia (less than 90 bpm) is associated with a high risk of miscarriage. Many women do not ovulate at around day 14, so findings after a single scan should always be interpreted with caution. The diagnosis of missed abortion is usually made by serial ultrasound scans demonstrating lack of gestational development.

If ultrasound scan demonstrates a 7mm embryo but cannot demonstrable a clearcut heartbeat, a missed abortion may be diagnosed. In such cases, it is reasonable to repeat the ultrasound scan in 7-10 days to avoid any error. The timing of a positive pregnancy test may also be helpful in this regard to assess the possible dates of conception. A positive pregnancy test 3 weeks previously would indicate a gestational age of at least 7 weeks.

Such information would be useful against the interpretation of the scans. In the presence of first trimester bleeding, ultrasonography is also indispensible in the early diagnosis of ectopic pregnancies and molar pregnancies.

DETERMINATION OF GESTATIONAL AGE AND ASSESSMENT OF FETAL SIZE

Fetal body measurements reflect the gestational age of the fetus. This is particularly true in early gestation. In patients with uncertain last menstrual periods, such measurements must be made as early as possible in pregnancy to arrive at a correct dating for the patient. In the latter part of pregnancy measuring body parameters will allow assessment of the size and growth of the fetus and will greatly assist in the diagnosis and management of intrauterine growth retardation (IUGR). The following measurements are usually made:

The Crown-rump Length (CRL)

This measurement can be made between 7 to 13 weeks and gives very accurate estimation of the gestational age. Dating with the CRL can be within 3-4 days of the last menstrual period. An important point to note is that when the due date has been set by an accurately measured CRL, it should not be changed by a subsequent scan. If another scan done 6 or 8 weeks later says that one should have a new due date which is further away, one should not normally change the date but should rather interpret the finding as that the baby is not growing at the expected rate.

The Biparietal Diameter (BPD)

The diameter between the 2 sides of the head. This is measured after 13 weeks. It increases from about 2.4 cm at 13 weeks to about 9.5 cm at term. Different babies of the same weight can have different head size, therefore dating in the later part of pregnancy is generally considered unreliable. Dating using the BPD should be done as early as is feasible.

The Femur Length (FL)

Measures the longest bone in the body and reflects the longitudinal growth of the fetus. Its usefulness is similar to the BPD. It increases from about 1.5 cm at 14 weeks to about 7.8 cm at term. Similar to the BPD, dating using the FL should be done as early as is feasible.

The Abdominal Circumference (AC)

The single most important measurement to make in late pregnancy. It reflects more of fetal size and weight rather than age. Serial measurements are useful in monitoring growth of the fetus. AC measurements should not be used for dating a fetus.

The weight of the fetus at any gestation can also be estimated with great accuracy using polynomial equations containing the BPD, FL, and AC. computer softwares and lookup charts are readily available. A BPD of 9.0 cm and an AC of 30.0 cm will give a weight estimate of 2.85 kg.

Diagnosis of Fetal Malformation

Many structural abnormalities in the fetus can be reliably diagnosed by an ultrasound scan, and these can usually be made before 20 weeks.

Common examples include hydrocephalus, anencephaly, myelomeningocoele, achondroplasia and other dwarfism, spina bifida, exomphalos, Gastroschisis, duodenal atresia and fetal hydrops.

With more recent equipment, conditions such as cleft lips/palate and congenital cardiac abnormalities are more readily diagnosed and at an earlier gestational age.

First trimester ultrasonic 'soft' markers for chromosomal abnormalities such as the absence of fetal nasal bone, an increased fetal nuchal translucency (the area at the back of the neck) are now in common use to enable detection of Down syndrome fetuses.

Placental Localization

Ultrasonography has become indispensible in the localization of the site of the placenta and determining its lower edges, thus making a diagnosis or an exclusion of placenta previa.

Other placental abnormalities in conditions such as diabetes, fetal hydrops, Rh isoimmunization and severe intrauterine growth retardation can also be assessed.

Multiple Pregnancies

In this situation, ultrasonography is invaluable in determining the number of fetuses, the chorionicity, fetal presentations, evidence of growth retardation and fetal anomaly, the presence of placenta previa, and any suggestion of twin-to-twin transfusion.

Hydramnios and Oligohydramnios

Excessive or decreased amount of liquor (amniotic fluid) can be clearly depicted by ultrasound. Both of these conditions can have adverse effects on the fetus. In both these situations, careful ultrasound examination should be made to exclude intraulterine growth retardation and congenital malformation in the fetus such as intestinal atresia, hydrops fetalis or renal dysplasia.

Doppler Ultrasound

The doppler shift principle has been used for a long time in fetal heart rate detectors. Further developments in doppler ultrasound technology in recent years have enabled a great expansion in its application in Obstetrics, particularly in the area of assessing and monitoring the well-being of the fetus, its progression in the face of intrauterine growth restriction, and the diagnosis of cardiac malformations. Doppler ultrasound is presently most widely employed in the detection of fetal cardiac pulsations and pulsations in the various fetal blood vessels.

The "Doptone" fetal pulse detector is a commonly used handheld device to detect fetal heartbeat using the same doppler principle. Blood flow characteristics in the fetal blood vessels can be assessed with Doppler 'flow velocity waveforms'. Diminished flow, particularly in the diastolic phase of a pulse cycle is associated with compromise in the fetus. Various ratios of the systolic to diastolic flow are used as a measure of this compromise.

The blood vessels commonly interrogated include the umbilical artery, the aorta, the middle cerebral arteries, the uterine arcuate arteries, and the inferior vena cava. The use of colour flow mapping can clearly depict the flow of blood in fetal blood vessels in a realtime scan, the direction of the flow being represented by different colours.

Colour doppler is particularly indispensible in the diagnosis of fetal cardiac and blood vessel defects, and in the assessment of the hemodynamic responses to fetal hypoxia and anemia. A more recent development is the Power Doppler

(Doppler angiography). It uses amplitude information from doppler signals rather than flow velocity information to visualize slow flow in smaller blood vessels.

A colour perfusion-like display of a particular organ such as the placenta overlapping on the 2-D image can be very nicely depicted. Doppler examinations can be performed abdominally and via the transvaginal route. The power emitted by a doppler device is greater than that used in a conventional 2-D scan.

Its use in early pregnancy is therefore cautioned. Doppler facilities are generally an integral part of modern ultrasound scanners. They merely would need to be switched on to function. One does not need to 'go' to another machine for the doppler investigations.

Transvaginal Scans

With specially designed probes, ultrasound scanning can be done with the probe placed in the vagina of the patient. This method usually provides better images in patients who are obese and/ or in the early stages of pregnancy. The better images are the result of the scanhead's closer proximity to the uterus and the higher frequency used in the transducer array resulting in higher resolving power. Fetal cardiac pulsation can be clearly observed as early as 6 weeks of gestation.

Vaginal scans are also becoming indispensible in the early diagnosis of ectopic pregnancies. An increasing number of fetal abnormalities are also being diagnosed in the first trimester using the vaginal scan. Transvaginal scans are also useful in the second trimester in the diagnosis of congenital anomalies.

3-D and 4-D Ultrasound

3-D ultrasound can furnish us with a 3 dimensional image of what we are scanning. The transducer takes a series of images, thin slices, of the subject, and the computer processes these images and presents them as a 3 dimensional image. Using computer controls, the operator can obtain views that might not be available using ordinary 2-D ultrasound scan. 3-dimensional ultrasound is quickly moving out of the research and development stages and is now widely employed in a clinical setting.

It too, is very much in the News. Faster and more advanced commercial models are coming into the market. The scans requires special probes and software to accumulate and render the images, and the rendering time has been reduced from minutes to fractions of a seconds. A good 3-D image is often very impressive to the parents. Further 2-D scans may be extracted from 3-D blocks of scanned information.

Volumetric measurements are more accurate and both doctors and parents can better appreciate a certain abnormality or the absence of a certain abnormality in a 3-D scan than a 2-D one and there is the possibility of increasing psychological bonding between the parents and the baby. An increasing volume of literature is accumulating on the usefulness of 3-D scans and the diagnosis of congenital anomalies could receive revived attention. Present evidence has already suggested that smaller defects such as spina bifida, cleft lips/palate, and polydactyl may be more lucidly demonstrated.

Other more subtle features such as low-set ears, facial dysmorphia or clubbing of feet can be better assessed, leading to more effective diagnosis of chromosomal abnormalities. The study of fetal cardiac malformations is also receiving attention. The ability to obtain a good 3-D picture is nevertheless still very much dependent on operator skill, the amount of liquor (amniotic fluid) around the fetus, its position and the degree of maternal obesity, so that a good image is not always readily obtainable. More recently, 4-D or dynamic 3-D scanners are in the market and the attraction of being able to look at the face and movements of your baby before birth was also enthusiastically reported in parenting and health magazines.

This is thought to have an important catalytic effect for mothers to bond to their babies before birth. What are known as 're-assurance scans' and the rather misnamed 'entertainment scans' have quickly become popular. Most experts do not consider that 3-D and 4-D ultrasound will be a mandatory evolution of our conventional 2-D scans, rather it is an additional piece of tool like doppler ultrasound. Most diagnosis will still be made with the 2-D scans. 3-D ultrasound appears to have great potential in research and in the study of fetal embryology. Whether 3-D ultrasound will provide unique information or merely supplemental information to the conventional 2-D scans will remain to be seen.

There is no hard and fast rule as to the number of scans a woman should have during her pregnancy. A scan is ordered when an abnormality is suspected on clinical grounds. Otherwise a scan is generally booked at about 7 weeks to confirm pregnancy, exclude ectopic or molar pregnancies, confirm cardiac pulsation and measure the crown-rump length for dating. A second scan is performed at 18 to 20 weeks mainly to look for congenital malformations, when the fetus is large enough for an accurate survey of the fetal anatomy. multiple pregnancies can be firmly diagnosed and dates and growth can also be assessed.

Placental position is also determined. Further scans may be necessary if abnormalities are suspected. Many centres are now performing an earlier screening scan at around 11-14 weeks to measure the fetal nuchal translucency and to evaluate the fetal nasal bone (and more recently, to detect tricuspid regurgitation) to aid in the diagnosis of Down Syndrome. Some centres will do blood test biochemical screening at the same visit. Further scans may sometimes be done at around 32 weeks or later to evaluate fetal size (to estimate the fetal weight) and assess fetal growth.

Or to follow up on possible abnormalities seen at an earlier scan. Placental position is further verified. The most common reason for having more scans in the later part of pregnancy is fetal growth retardation. Doppler scans may also be necessary in that situation. The total number of scans will vary depending on whether a previous scan has detected certain abnormalities that require follow-up assessment. What is often referred to as a Level II scan merely indicates a "targeted" examination where it is done when an indication is present or when an abnormality is suspected in a previous examination.

In fact professional bodies such as the American Institute of Ultrasound in Medicine does not endorse or encourage the use of these terms. A more "thorough" examination is usually done at an a perinatal centre or specialised clinic where more expertise and better equipments may be present. One should not dwell too much on the definitions or guidelines for a level II ultrasound scan. The prenatal sonologist should always try very hard to look for and assess any abnormality that may be present in the fetus.

It is not very meaningful to be talking about level III or even level IV scans. That a pregnancy should be scanned at 18 to 20 weeks as a rule is gradually becoming a matter of routine practice. A rather thorough discussion paper on Ultrasound screening in pregnancy can be found here. Read also the RCOG's paper on routine screening in pregnancy.

It has been over 40 years since ultrasound was first used on pregnant women. Unlike X-rays, ionizing irradiation is not present and embryotoxic effects associated with such irradiation should not be relevant. The use of high intensity ultrasound is associated with the effects of "cavitation" and "heating" which can be present with prolonged insonation in laboratory situations. Although certain harmful effects in cells are observed in a laboratory setting, abnormalities in embryos and offsprings of animals and humans have not been unequivocally demonstrated in the large amount of studies that have so far appeared in the medical literature purporting to the use of diagnostic ultrasound in the clinical setting.

Apparent ill-effects such as low birthweight, speech and hearing problems, brain damage and non-right-handedness reported in small studies have not been confirmed or substantiated in larger studies from Europe. The complexity of some of the studies have made the observations difficult to interpret. Every now and then ill effects of ultrasound on the fetus appears as a news item in papers and magazines. Continuous vigilance is necessary particularly in areas of concern such as the use of pulsed Doppler in the first trimester.

The greatest risks arising from the use of ultrasound are the possible over- and under- diagnosis brought about by inadequately trained staff, often working in relative isolation and using poor equipment. A discussion on the various possible effects of ultrasound on the human fetus can be found here.

Ultrasound scans should best be performed when there is a clear indication to do so. When there is, safety considerations should not be an issue to prevent its prudent use. It should be bornt in mind that prenatal ultrasound cannot diagnose all malformations and problems of an unborn baby, so one should never interpret a normal scan report as a guarantee that the baby will be completely normal. Some abnormalities are very difficult to find or to be absolutely certain about. Some conditions, like hydrocephalus, may not have been obvious at the time of the earlier scan.

The position of the baby in the uterus has a great deal to do with how well one sees certain organs such as the heart, face and spine. Sometimes a repeat examination has to be scheduled the following day, in the hopes the baby has moved. Images tend also to be strikingly clear in skinny patients with lots of amniotic fluid, and frustratingly fuzzy in obese women, particularly if there is not much amniotic fluid as in cases of growth restriction.

As in almost every endeavor, there is also a wide difference in the skill, training, talent, and interest of the sonographer or sonologists. The improvements in equipment has also lead to the earlier detection of abnormal structures in the fetus bringing along with it *"false positives"* and *"difficult-to-be-sure-what-will-happen"* diagnosis that could generate huge amount of undue anxiety in patients.

Chapter 10

Computed Tomography

Computed tomography (CT), was originally known as "EMI scan" as it was developed at a research branch of EMI, a company best known today for its music and recording business. It was later known as *computed axial tomography* (CAT or CT scan) and *body section roentgenography.*

Computed tomography is a medical imaging method employing tomography where digital geometry processing is used to generate a three-dimensional image of the internals of an object from a large series of two-dimensional X-ray images taken around a single axis of rotation. The word "tomography" is derived from the Greek *tomos* (slice) and *graphein* (to write).

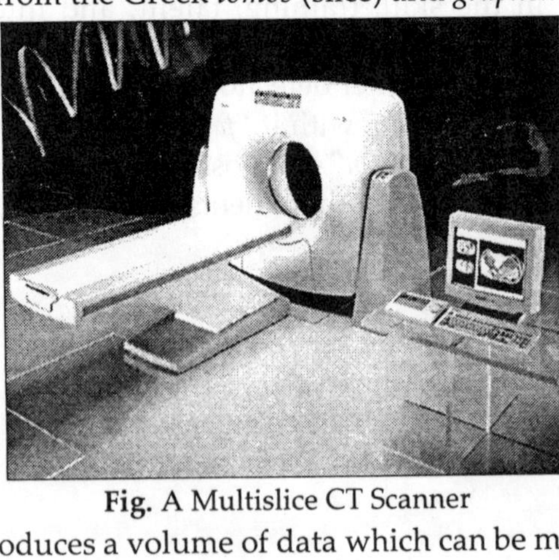

Fig. A Multislice CT Scanner

CT produces a volume of data which can be manipulated, through a process known as *windowing,* in order to demonstrate various structures based on their ability to block the X-ray beam. Although historically the images generated were in the axial or transverse plane (orthogonal to the long axis of the body), modern scanners allow this volume of data to be reformatted in various planes or even as volumetric (3D) representations of structures.

Although most common in healthcare, CT is also used in other fields, nondestructive materials testing. Another example is the DigiMorph project at the University of Texas at Austin which uses a CT scanner to study biological and paleontological specimens. In the early 1930s the Italian radiologist Alessandro Vallebona proposed a method to represent a single slice of the body on the radiographic film. This exam was known as tomography.

The idea is based on simple principles of projective geometry: moving synchronously and in opposite directions the X-ray tube and the film, which are connected together by a rod whose pivot point is the focus; the image created by the points on the focal plane appears sharper, while the images of the other points annihilate as noise. This is only marginally effective, as blurring occurs only in the "x" plane. There are also more complex devices which can move in more than one plane and perform more effective blurring.

Tomography has been one of the pillars of radiologic diagnostics until the late 1970s, when the availability of minicomputers and of the transverse axial scanning method, this last due to Godfrey Newbold Hounsfield and Allan McLeod Cormack, gradually supplanted it as the modality of CT.

The first commercially viable CT scanner was invented by Sir Godfrey Newbold Hounsfield in Hayes, United Kingdom at EMI Central Research Laboratories using X-rays. Hounsfield conceived his idea in 1967, and it was publicly announced in 1972. Allan McLeod Cormack of Tufts University, Massachusetts, USA independently invented a similar process, and both Hounsfield and Cormack shared the 1979 Nobel Prize in Medicine.

The original 1971 prototype took 160 parallel readings through 180 angles, each 1° apart, with each scan taking a little over five minutes. The images from these scans took 2.5 hours to be processed by algebraic reconstruction techniques on a large computer.

The scanner had a single photomultiplier detector and operated on the Translate/Rotate principle. The CT scanner was "the greatest legacy" of The Beatles, with the massive profits resulting from their record sales enabling EMI to fund scientific research, including into computerised tomography. The first production X-ray CT machine (in fact called the "EMI-Scanner") was limited to making tomographic sections of the brain, but acquired the image data in about 4 minutes (scanning two adjacent slices) and the computation time (using a Data General Nova minicomputer) was about 7 minutes per picture.

This scanner required the use of a water-filled Perspex tank with a pre-shaped rubber "head-cap" at the front, which enclosed the patient's head. The water-tank was used to reduce the dynamic range of the radiation reaching the detectors (between scanning outside the head compared with scanning through the bone of the skull).

The images were relatively low resolution, being composed of a matrix of only 80 x 80 pixels. The first EMI-Scanner was installed in Atkinson Morley's Hospital in Wimbledon, England, and the first patient brain-scan was made with it in 1972. In the U.S., the first installation was at the Mayo Clinic.

As a tribute to the impact of this system on medical imaging the Mayo Clinic has an EMI scanner on display in the Radiology Department. The first CT system that could make images of any part of the body, and did not require the "water tank" was the ACTA (Automatic Computerized Transverse Axial) scanner designed by Robert S. Ledley, DDS at Georgetown University. This machine had 30 photomultiplier tubes as detectors and completed a scan in only 9 translate/rotate cycles, much faster than the EMI-scanner.

It used a DEC PDP11/34 minicomputer both to operate the servo-mechanisms and to acquire and process the images. The Pfizer drug company acquired the prototype from the university, along with rights to manufacture it. Pfizer then began making copies of the prototype, calling it the "200FS" (FS meaning Fast Scan), which were selling as fast as they could make them. This unit produced images in a 256x256 matrix, with much better definition than the EMI-Scanner's 80x80.

Tomosynthesis

Digital tomosynthesis combines digital image capture and processing with simple tube/detector motion as used in conventional radiographic tomography - although there are some similarities to CT, it is a separate technique. In CT, the source/detector makes a complete 360 degree rotation about the subject obtaining a complete set of data from which images may be reconstructed.

In digital tomosynthesis, only a small rotation angle (e.g. 40 degrees) with a small number of discrete exposures (e.g. 10) are used. This incomplete set of data can be digitally processed to yield images similar to conventional tomography with a limited depth of field. However, because the image processing is digital, a series of slices at different depths and with different thicknesses can be reconstructed from the same acquisition, saving both time and radiation exposure.

Because the data acquired is incomplete, tomosynthesis is unable to offer the extremely narrow slice widths that CT offers. However, higher resolution detectors can be used, allowing very-high in-plane resolution, even if the Z-axis resolution is poor. The primary interest in tomosynthesis is in breast imaging, as an extension to mammography, where it may offer better detection rates, with little extra increase in radiation exposure.

Reconstruction algorithms for tomosynthesis are significantly different from conventional CT, as the conventional filtered back projection algorithm requires a complete set of data.

Iterative algorithms based upon expectation maximization are most commonly used, but are extremely computationally intensive. Some manufacturers have produced practical systems using commercial GPUs to perform the reconstruction.

Tomography

CT's primary benefit is the ability to separate anatomical structures at different depths within the body. A form of tomography can be performed by moving the X-ray source and detector during an exposure.

Anatomy at the target level remains sharp, while structures at different levels are blurred. By varying the extent and path of motion, a variety of effects can be obtained, with variable depth of field and different degrees of blurring of 'out of plane' structures.

Although largely obsolete, conventional tomography is still used in specific situations such as dental imaging (orthopantomography) or in intravenous urography.

MODERN CT ACQUISITION

Electron Beam CT

Electron beam tomography (EBCT) was introduced in the early 1980s, by medical physicist Andrew Castagnini, as a method of improving the temporal resolution of CT scanners. Because the X-ray source has to rotate by over 180 degrees in order to capture an image the technique is inherently unable to capture dynamic events or movements that are quicker than the rotation time.

Instead of rotating a conventional X-ray tube around the patient, the EBCT machine houses a huge vacuum tube in which an electron beam is electro-magnetically steered towards an array of tungsten X-ray anodes arranged circularly around the patient. Each anode is hit in turn by the electron beam and emits X-rays that are collimated and detected as in conventional CT. The lack of moving parts allows very quick scanning, with single slice acquisition in 50-100 ms, making the technique ideal for capturing images of the heart. EBCT has found particular use for assessment of coronary artery calcium, a means of predicting risk of coronary artery disease.

The very high cost of EBCT equipment, and its poor flexibility (EBCT scanners are essentially single-purpose cardiac scanners), has led to poor uptake; fewer than 150 of these scanners have been installed worldwide. EBCT's role in cardiac imaging is rapidly being supplanted by high-speed multi-detector CT, which can achieve near-equivalent temporal resolution with much faster z-axis coverage.

Helical or Spiral CT

Helical, also called spiral, CT was introduced in the early 1990s, with much of the development led by Willi Kalender and Kazuhiro Katada. In older CT scanners, the X-ray source would move in a circular fashion to acquire a single 'slice', once the slice had been completed, the scanner table would move to position the patient for the next slice; meanwhile the X-ray source/detectors would reverse direction to avoid tangling their cables.

In helical CT the X-ray source (and detectors in 3rd generation designs) are attached to a freely rotating gantry. During a scan, the table moves the patient smoothly through the scanner; the name derives from the helical path traced out by the X-ray beam. It was the development of two technologies that made helical CT practical: slip rings to transfer power and data on and off the rotating gantry, and the switched mode power supply powerful enough to supply the X-ray tube, but small enough to be installed on the gantry.

The major advantage of helical scanning compared to the traditional shoot-and-step approach, is speed; a large volume can be covered in 20-60 seconds. This is advantageous for a number or reasons: 1) often the patient can hold their breath for the entire study, reducing motion artifacts, 2) it allows for more optimal use of intravenous contrast enhancement, and 3) the study is quicker than the equivalent conventional CT permitting the use of higher resolution acquisitions in the same study time. The data obtained from spiral CT is often well-suited for 3D imaging because of the lack of motion mis-registration and the increased out of plane resolution. These major advantages led to the rapid rise of helical CT as the most popular type of CT technology.

Despite the advantages of helical scanning, there are a few circumstances where it may not be desirable - there is, of course, no difficulty in configuring a helical capable scanner for scanning in shoot-and-step mode.

All other factors being equal, helical CT has slightly lower z-axis resolution than step-and-shoot (due to the continual movement of the patient). Where z-resolution is critical but where it is undesirable to scan at a higher resolution setting (due to the higher radiation exposure required) e.g. brain imaging, step-and-shoot may still be the preferred method.

Multislice CT

Multislice CT scanners are similar in concept to the helical or spiral CT but there are more than one detector ring. It began with two rings in mid nineties, with a 2 solid state ring model designed and built by Elscint (Haifa) called CT TWIN, with one second rotation (1993): It was followed by other manufacturers.

Later, it was presented 4, 8, 16, 32, 40 and 64 detector rings, with increasing rotation speeds. Current models have up to 3 rotations per second, and isotropic resolution of 0.35mm voxels with z-axis scan speed of up to 18 cm/s.

This resolution exceeds that of High Resolution CT techniques with single-slice scanners, yet it is practical to scan adjacent, or overlapping, slices - however, image noise and radiation exposure significantly limit the use of such resolutions.

The major benefit of multi-slice CT is the increased speed of volume coverage. This allows large volumes to be scanned at the optimal time following intravenous contrast administration; this has particularly benefitted CT angiography techniques - which rely heavily on precise timing to ensure good demonstration of arteries.

Computer power permits increasing the postprocessing capabilities on workstations. Bone suppression, volume rendering in real time, with a natural visualization of internal organs and structures, and automated volume reconstruction really change the way diagnostic is performed on CT studies and this models become true volumetric scanners.

The ability of multi-slice scanners to achieve isotropic

resolution even on routine studies means that maximum image quality is not restricted to images in the axial plane - and studies can be freely viewed in any desired plane.

Dual Source CT

Siemens introduced a CT model with dual X-ray tube and dual array of 64 slice detectors, at the 2005 Radiological Society of North America (RSNA) medical meeting. Dual sources increase the temporal resolution by reducing the rotation angle required to acquire a complete image, thus permitting cardiac studies without the use of heart rate lowering medication, as well as permitting imaging of the heart in systole.

The use of two x-ray units makes possible the use of dual energy imaging, which allows an estimate of the average atomic number in a voxel, as well as the total attenutaion. This permits automatic differentiation of calcium (e.g. in bone, or diseased arteries) from iodine (in contrast medium) or titanium (in stents) - which might otherwise be impossible to differentiate. It may also improve the characterization of tissues allowing better tumor differentiation.

Topogram

A Scout image is used in planning the exam and to establish where the target organs are located. The beginning and end of the scan are set by the target region and the location of the patient on the table.

Once the Scout image is created it is used to determine the extent of the desired Axial/Helical scan. During the Scout scan the gantry is rotated to a fixed position and the table is translated as x-ray is delivered. The image appears similar to a radiograph.

Axial

In axial "step and shoot" acquisitions each slice/volume is taken and then the table is incremented to the next location. In multislice scanners each location is multiple slices and represents a volume of the patient anatomy. Tomographic reconstruction is used to generate Axial images.

Cine

A cine acquisition is used when the temporal nature is important. This is used in Perfusion applications to evaluate blood flow, blood volume and mean transit time. Cine is a time sequence of axial images. In a Cine acquisition the cradle is stationary and the gantry rotates continuously. Xray is delivered at a specified interval and duration.

Helical/Spiral

Helical is a very fast way to examine the target anatomy. The volume is scanned very quickly because the table is in constant motion as the gantry rotates continuously. There is no interscan delay between slices as in a Axial acquisition.

DRR

A Digitally Reconstructed Radiograph is a simulation of a conventional 2D x-ray image, created from computed tomography (CT) data. A radiograph, or conventional x-ray image, is a single 2D view of total x-ray absorption through the body along a given axis. Two objects in front of one another will overlap in the image.

By contrast, a 3D CT image gives a volumetric representation. Sometimes one must compare CT data to a classical radiograph, and this can be done by comparing a DRR based on the CT data. An early example of their use is the beam's eye view (BEV) as used in radiotherapy planning. In this application, a BEV is created for a specific patient and is used to help plan the treatment.

DRRs are created by summing CT intensities along a ray from each pixel to the simulated x-ray source.

Since 1993, the Visible Human Project (VHP) has made full body CT data available to researchers. This has allowed several universities and commercial companies to try and create DRR's. These have been suggested as useful for training simulations in Radiology and Diagnostic Radiography. It takes a significant number of calculations to create a summative 2D image from a large amount of 3D data. This is an area of medical science and education that has benefited from the advancing of graphics card technology, driven by the computer games industry.

Another novel use of DRR's is in identification of the dead from old radiographic records, by comparing them to DRR's created from CT data.

Synchrotron X-ray Tomographic Microscopy

Synchrotron X-ray tomographic microscopy is a 3-D scanning technique that allows non-invasive high definition scans of objects with details as fine as 1,000th of a millimetre, meaning it has two to three thousand times the resolution of a traditional medical CT scan.

Synchrotron X-ray tomographic microscopy has been applied in the field of palaeontology to perform non-destructive internal examination of fossils, including fossil embryos to be made. Scientists feel this technology has the potential to revolutionize the field of paleontology. The first team to use the technique have published their findings in Nature, which they believe "could roll back the evolutionary history of arthropods like insects and spiders."

Archaeologists are increasingly turning to Synchrotron X-ray tomographic microscopy as a non-destructive means to examine ancient specimens.

X-ray Tomography

X-ray Tomography is a branch of X-ray microscopy. A series of projection images are used to calculate a three dimensional reconstruction of an object. The technique has found many applications in materials science and later in biology and biomedical research. In terms of the latter, the National Centre for X-ray Tomography (NCXT) is one of the principal developers of this technology, in particular for imaging whole, hydrated cells.

256+ Slice CT

At RSNA 2007, Philips announced a 256 slice scanner, while Toshiba announced a "dynamic volume" scanner based on 320 slices. The majority of published data with regard to both technical and clinical aspects of the systems have been related to the prototype unit made by Toshiba Medical Systems.

The recent 3 month Beta installation at Johns Hopkins Press Release using a Toshiba system tested the clinical capabilities of this technology JHU Gazette. The technology currently remains in a development phase but has demonstrated the potential to significantly reduce radiation exposure by eliminating the requirement for a helical examination in both cardiac CT angiography and whole brain perfusion studies for the evaluation of stroke.

Inverse Geometry CT

Inverse geometry CT (IGCT) is a novel concept which is being investigated as refinement of the classic third generation CT design. Although the technique has been demonstrated on a laboratory proof-of-concept device, it remains to be seen whether IGCT is feasible for a practical scanner. IGCT reverses the shapes of the detector and X-ray sources.

The conventional third-generation CT geometry uses a point source of X-rays, which diverge in a fan beam to act on a linear array of detectors. In multidetector computed tomography (MDCT), this is extended in 3 dimensions to a conical beam acting on a 2D array of detectors. The IGCT concept, conversely, uses an array of highly collimated X-ray sources which act on a point detector. By using a principle similar to electron beam tomography (EBCT), the individual sources can be activated in turn by steering an electron beam onto each source target.

The rationale behind IGCT is that it avoids the disadvantages of the cone-beam geometry of third generation MDCT. As the z-axis width of the cone beam increases, the quantity of scattered radiation reaching the detector also increases, and the z-axis resolution is thereby degraded - because of the increasing z-axis distance that each ray must traverse.

This reversal of roles has extremely high intrinsic resistance to scatter; and, by reducing the number of detectors required per slice, it makes the use of better performing detectors (e.g. ultra-fast photon counting detectors) more practical. Because a separate detector can be used for each 'slice' of sources, the conical geometry can be replaced with an array of fans, permitting z-axis resolution to be preserved.

DIAGNOSTIC USE

Since its introduction in the 1970s, CT has become an important tool in medical imaging to supplement X-rays and medical ultrasonography. Although it is still quite expensive, it is the gold standard in the diagnosis of a large number of different disease entities.

It has more recently begun to also be used for preventive medicine or screening for disease, CT colonography for patients with a high risk of colon cancer. Although a number of institutions offer full-body scans for the general population, this practice remains controversial due to its lack of proven benefit, cost, radiation exposure, and the risk of finding 'incidental' abnormalities that may trigger additional investigations.

Cranial

Diagnosis of *cerebrovascular accidents* and *intracranial hemorrhage* is the most frequent reason for a "head CT" or "CT brain". Scanning is done with or without intravenous contrast agents. CT generally does not exclude infarct in the acute stage of a stroke.

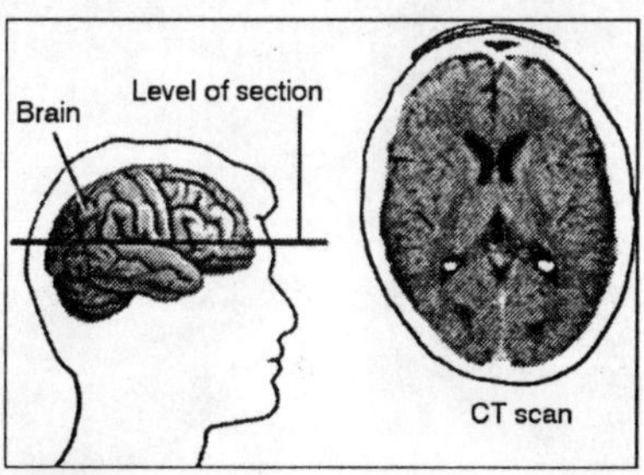

Fig. Normal CT scan of the Head

However, CT is useful to exclude an intra-cranial hemorrhage as a cause, or complication of the stroke. For the detection of acute hemorrhage, especially subarachnoid hemorrhage, CT is the test of choice as it is more sensitive than MRI.

For detection of *tumors*, CT scanning with IV contrast is occasionally used but is less sensitive than magnetic resonance imaging (MRI). CT has an important role in evaluation of the functioning of a ventriculoperitoneal shunt by demonstrating the volume of the ventricular system.

Although CT cannot assess *intracranial pressure*, it can demonstrate several important causes of raised intracranial pressure. Despite the limitation that CT may not detect raised intracranial pressure, it may help in the clinical decision to perform lumbar puncture and is often performed in this context.

CT is also useful in the setting of trauma for evaluating facial and skull *fractures*.

In the head/neck/mouth area, CT scanning is used for surgical planning for craniofacial and dentofacial deformities, evaluation of cysts and some tumors of the jaws/paranasal sinuses/nasal cavity/orbits, diagnosis of the causes of chronic sinusitis, and for planning of dental implant reconstruction.

Chest

CT is excellent for detecting both acute and chronic changes in the lung parenchyma.

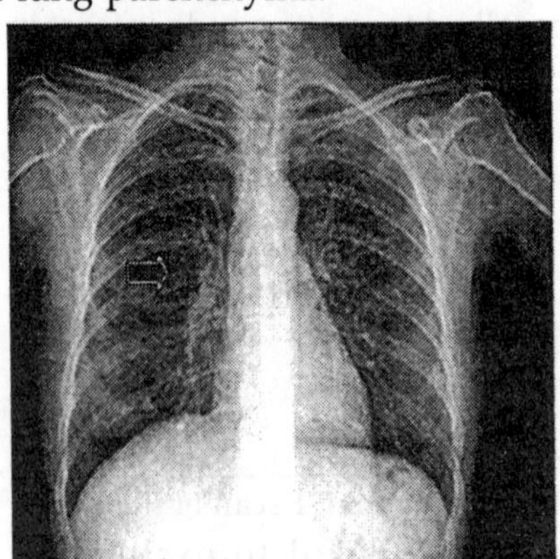

Fig. CT scan of Chest

A variety of different techniques are used depending on the suspected abnormality. For evaluation of chronic interstitial processes (emphysema, fibrosis, and so forth), thin sections with high spatial frequency reconstructions are used - often scans are performed both in inspiration and expiration. This special technique is called High resolution CT (HRCT).

For detection of airspace disease (such as pneumonia) or cancer, relatively thick sections and general purpose image reconstruction techniques may be adequate. IV contrast may also be used as it clarifies the anatomy and boundaries of the great vessels and improves assessment of the mediastinum and hilar regions for lymphadenopathy; this is particularly important for accurate assessment of cancer.

CT angiography of the chest is also becoming the primary method for detecting pulmonary embolism (PE) and aortic dissection, and requires accurately timed rapid injections of contrast (Bolus Tracking) and high-speed helical scanners. CT is the standard method of evaluating abnormalities seen on chest X-ray and of following findings of uncertain acute significance.

PULMONARY ANGIOGRAM

CT pulmonary angiogram (CTPA) is a medical diagnostic test used to diagnose pulmonary embolism (PE). It employs computed tomography to obtain an image of the pulmonary arteries.

Diagnostic Use

It is a preferred choice of imaging in the diagnosis of PE due to its minimally invasive nature for the patient, whose only requirement for the scan is a cannula (usually a 20G). Before this test is requested, it is usual for the referring clinician to have carried out a D-dimer blood test and requested a chest X-Ray to rule out any other possible differential diagnosis.

Abdominal and Pelvic

CT is a sensitive method for diagnosis of abdominal diseases. It is used frequently to determine stage of cancer and to follow progress. It is also a useful test to investigate acute abdominal pain.

Renal/urinary stones, appendicitis, pancreatitis, diverticulitis, abdominal aortic aneurysm, and bowel obstruction are conditions that are readily diagnosed and assessed with CT. CT is also the first line for detecting solid organ injury after trauma. Oral and/or rectal contrast may be used depending on the indications for the scan. A dilute (2% w/v) suspension of barium sulfate is most commonly used. The concentrated barium sulfate preparations used for fluoroscopy e.g. barium enema are too dense and cause severe artifacts on CT.

Iodinated contrast agents may be used if barium is contraindicated (e.g. suspicion of bowel injury). Other agents may be required to optimize the imaging of specific organs: e.g. rectally administered gas (air or carbon dioxide) for a colon study, or oral water for a stomach study. CT has limited application in the evaluation of the *pelvis*. For the female pelvis in particular, ultrasound and MRI are the imaging modalities of choice. Nevertheless, it may be part of abdominal scanning (e.g. for tumors), and has uses in assessing fractures. CT is also used in osteoporosis studies and research along side DXA scanning.

Both CT and DXA can be used to assess bone mineral density (BMD) which is used to indicate bone strength, however CT results do not correlate exactly with DXA (the gold standard of BMD measurement). CT is far more expensive, and subjects patients to much higher levels of ionizing radiation, so it is used infrequently.

Interpretation

A normal CTPA scan will show the contrast filling the pulmonary vessels, looking bright white. Ideally the aorta should be empty of contrast, to reduce any partial volume artefact which may result in a false positive. Any mass filling defects, such as an embolus, will appear dark in place of the contrast, filling / blocking the space where blood should be flowing into the lungs.

Cardiac

With the advent of subsecond rotation combined with multi-slice CT (up to 64-slice), high resolution and high speed can be obtained at the same time, allowing excellent imaging of the coronary arteries (cardiac CT angiography). Images with an even higher temporal resolution can be formed using retrospective ECG gating. In this technique, each portion of the heart is imaged more than once while an ECG trace is recorded.

The ECG is then used to correlate the CT data with their corresponding phases of cardiac contraction. Once this correlation is complete, all data that were recorded while the heart was in motion (systole) can be ignored and images can be made from the remaining data that happened to be acquired while the heart was at rest (diastole). In this way, individual frames in a cardiac CT investigation have a better temporal resolution than the shortest tube rotation time.

Because the heart is effectively imaged more than once, cardiac CT angiography results in a relatively high radiation exposure around 12 mSv. For the sake of comparison, a chest X-ray carries a dose of approximately 0.02 to 0.2 mSv and natural background radiation exposure is around 0.01 mSv/day. Thus, cardiac CTA is equivalent to approximately 100-600 chest X-rays or over 3 years worth of natural background radiation. Methods are available to decrease this exposure, however, such as prospectively decreasing radiation output based on the concurrently acquired ECG (aka tube current modulation.)

This can result in a significant decrease in radiation exposure, at the risk of compromising image quality if there is any arrhythmia during the acquisition. The significance of radiation doses in the diagnostic imaging range has not been proven, although the possibility of inducing an increased cancer risk across a population is a source of significant concern.

This potential risk must be weighed against the competing risk of not performing a test and potentially not diagnosing a significant health problem such as coronary artery disease.

It is uncertain whether this modality will replace invasive coronary catheterization. Currently, it appears that the greatest utility of cardiac CT lies in ruling out coronary artery disease rather than ruling it in.

This is because the test has a high sensitivity (greater than 90%) and thus a negative test result means that a patient is very unlikely to have coronary artery disease and can be worked up for other causes of their chest symptoms. This is termed a high negative predictive value. A positive result is less conclusive and often will be confirmed (and possibly treated) with subsequent invasive angiography.

For the record, the positive predictive value of cardiac CTA is estimated at approximately 82% and the negative predictive value is around 93%. Dual Source CT scanners, introduced in 2005, allow higher temporal resolution by acquiring a full CT slice in only half a rotation, thus reducing motion blurring at high heart rates and potentially allowing for shorter breath-hold time.

This is particularly useful for ill patients who have difficulty holding their breath or who are unable to take heart-rate lowering medication.

The speed advantages of 64-slice MSCT have rapidly established it as the minimum standard for newly installed CT scanners intended for cardiac scanning.

Manufacturers are now actively developing 256-slice and true 'volumetric' scanners, primarily for their improved cardiac scanning performance. The latest MSCT scanners acquire images only at 70-80% of the R-R interval (late diastole).

This prospective gating can reduce effective dose from 10-15mSv to as little as 1.2mSv in follow-up patients acquiring at 75% of the R-R interval. Effective doses at a centre with well trained staff doing coronary imaging can average less than the doses for conventional coronary angiography. CT is often used to image complex fractures, especially ones around joints, because of its ability to reconstruct the area of interest in multiple planes. Fractures, ligamentous injuries and dislocations can easily be recognised with a 0.2 mm resolution.

ADVANTAGES AND HAZARDS

RADIATION EXPOSURE

CT is regarded as a moderate to high radiation diagnostic technique. While technical advances have improved radiation efficiency, there has been simultaneous pressure to obtain higher-resolution imaging and use more complex scan techniques, both of which require higher doses of radiation. The improved resolution of CT has permitted the development of new investigations, which may have advantages; e.g. Compared to conventional angiography, CT angiography avoids the invasive insertion of an arterial catheter and guidewire; CT colonography may be as useful as a barium enema for detection of tumors, but may use a lower radiation dose. The greatly increased availability of CT, together with its value for an increasing number of conditions, has been responsible for a large rise in popularity.

So large has been this rise that, in the most recent comprehensive survey in the UK, CT scans constituted 7% of all radiologic examinations, but contributed 47% of the total collective dose from medical X-ray examinations in 2000/2001. Increased CT usage has led to an overall rise in the total amount of medical radiation used, despite reductions in other areas. The radiation dose for a particular study depends on multiple factors: volume scanned, patient build, number and type of scan sequences, and desired resolution and image quality. Additionally, two helical CT scanning parameters that can be adjusted easily and that have a profound effect on radiation dose are tube current and pitch.

CT scans of children have been estimated to produce non-negligible increases in the probability of lifetime cancer mortality leading to calls for the use of reduced current settings for CT scans of children. A 2007 report in the New England Journal of Medicine suggested that the radiation from current CT-scan use may cause as many as 1 in 50 future cases of cancer. An average CT scan can expose a patient to between 1,000 to 10,000 millirems of radiation, depending on the exact machine and the examination being performed. However,

Japanese people who were 1 mile from ground zero received only 3,ooo millirems of radiation, on average.

First, CT completely eliminates the superimposition of images of structures outside the area of interest. Second, because of the inherent high-contrast resolution of CT, differences between tissues that differ in physical density by less than 1% can be distinguished. Third, data from a single CT imaging procedure consisting of either multiple contiguous or one helical scan can be viewed as images in the axial, coronal, or sagittal planes, depending on the diagnostic task. This is referred to as multiplanar reformatted imaging.

Adverse Effect

Because CT scans rely on intravenously administered contrast agents in order to provide superior image quality, there is a low but non-negligible level of risk associated with the contrast agents themselves. Certain patients may experience severe and potentially life-threatening allergic reactions to the contrast dye. The contrast agent may also induce kidney damage. The risk of this is increased with patients who have preexisting renal insufficiency, preexisting diabetes, or reduced intravascular volume.

In general, if a patient has normal kidney function, then the risks of contrast nephropathy are negligible. Patients with mild kidney impairment are usually advised to ensure full hydration for several hours before and after the injection. For moderate kidney failure, the use of iodinated contrast should be avoided; this may mean using an alternative technique instead of CT e.g. MRI. Perhaps paradoxically, patients with severe renal failure requiring dialysis do not require special precautions, as their kidneys have so little function remaining that any further damage would not be noticeable and the dialysis will remove the contrast agent.

Process

X-ray slice data is generated using an X-ray source that rotates around the object; X-ray sensors are positioned on the opposite side of the circle from the X-ray source. The earliest sensors were scintillation detectors, with photomultiplier tubes excited by (typically) sodium iodide crystals. Modern detectors use the ionization principle and are filled with low-pressure Xenon gas.

Many data scans are progressively taken as the object is gradually passed through the gantry. They are combined together by the mathematical procedures known as tomographic reconstruction. The data are arranged in a matrix in memory, and each data point is convolved with its neighbours according with a seed algorithm using Fast Fourier Transform techniques.

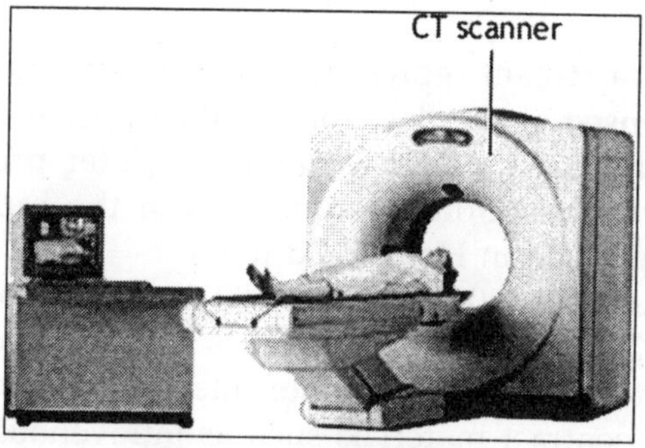

Fig. CT scan Illustration

This dramatically increases the resolution of each Voxel (volume element). Then a process known as Back Projection essentially reverses the acquisition geometry and stores the result in another memory array.

This data can then be displayed, photographed, or used as input for further processing, such as multi-planar reconstruction. Newer machines with faster computer systems and newer software strategies can process not only individual cross sections but continuously changing cross sections as the gantry, with the object to be imaged, is slowly and smoothly slid through the X-ray circle. These are called *helical* or *spiral CT* machines.

Their computer systems integrate the data of the moving individual slices to generate three dimensional volumetric information (3D-CT scan), in turn viewable from multiple different perspectives on attached CT workstation monitors.

This type of data acquisition requires enormous processing power, as the data are arriving in a continuous stream and must be processed in real-time.

In conventional CT machines, an X-ray tube and detector are physically rotated behind a circular shroud; in the electron beam tomography (EBT) the tube is far larger and higher power to support the high temporal resolution.

The electron beam is deflected in a hollow funnel shaped vacuum chamber. X-rays are generated when the beam hits the stationary target. The detector is also stationary. This arrangement can result in very fast scans, but is extremely expensive.

The data stream representing the varying radiographic intensity sensed at the detectors on the opposite side of the circle during each sweep is then computer processed to calculate cross-sectional estimations of the radiographic density, expressed in Hounsfield units.

Sweeps cover 360 or just over 180 degrees in conventional machines, 220 degrees in EBT. CT is used in medicine as a diagnostic tool and as a guide for interventional procedures. Sometimes contrast materials such as intravenous iodinated contrast are used. This is useful to highlight structures such as blood vessels that otherwise would be difficult to delineate from their surroundings. Using contrast material can also help to obtain functional information about tissues. Pixels in an image obtained by CT scanning are displayed in terms of relative radiodensity.

The pixel itself is displayed according to the mean attenuation of the tissue(s) that it corresponds to on a scale from -1024 to +3071 on the Hounsfield scale. Pixel is a two dimensional unit based on the matrix size and the field of view. When the CT slice thickness is also factored in, the unit is known as a Voxel, which is a three dimensional unit.

The phenomenon that one part of the detector cannot differ between different tissues is called the *"Partial Volume Effect"*. That means that a big amount of cartilage and a thin layer of compact bone can cause the same attenuation in a voxel as hyperdense cartilage alone.

Water has an attenuation of 0 Hounsfield units (HU) while air is -1000 HU, cancellous bone is typically +400 HU, cranial bone can reach 2000 HU or more (os temporale) and can cause artifacts.

The attenuation of metallic implants depends on atomic number of the element used: Titanium usually has an amount of +1000 HU, iron steel can completely extinguish the X-ray and is therefore responsible for well-known line-artifacts in computed tomograms.

Windowing

Windowing is the process of using the calculated Hounsfield units to make an image. The display device, as well as the human eye, can only resolve 256 shades of gray. These shades of gray can be distributed over a wide range of HU values to get an overview of structures that attenuate the beam to widely varying degrees. Alternatively, these shades of gray can be distributed over a narrow range of HU values centreed over the average HU value of a particular structure to be evaluated.

In this way, subtle variations in the internal makeup of the structure can be discerned. This is a commonly used image processing technique known as contrast compression. To evaluate the abdomen in order to find subtle masses in the liver, one might use liver windows. Choosing 70 HU as an average HU value for liver, the shades of gray can be distributed over a narrow window or range. One could use 170 HU as the narrow window, with 85 HU above the 70 HU average value; 85 HU below it.

Therefore the liver window would extend from -15 HU to +155 HU. All the shades of gray for the image would be distributed in this range of Hounsfield values. Any HU value below -15 would be pure black, and any HU value above 155 HU would be pure white in this example. Using this same logic, bone windows would use a *"wide window"* (to evaluate everything from fat-containing medullary bone that contains the marrow, to the dense cortical bone), and the centre or level would be a value in the hundreds of Hounsfield units.

3D Rendering Techniques

Surface Rendering

A threshold value of radiodensity is chosen by the operator (e.g. a level that corresponds to bone). A threshold level is set, using edge detection image processing algorithms. From this, a 3-dimensional model can be constructed and displayed on screen. Multiple models can be constructed from various different thresholds, allowing different colours to represent each anatomical component such as bone, muscle, and cartilage. However, the interior structure of each element is not visible in this mode of operation.

Volume Rendering

Surface rendering is limited in that it will only display surfaces which meet a threshold density, and will only display the surface that is closest to the imaginary viewer. In volume rendering, transparency and colours are used to allow a better representation of the volume to be shown in a single image - e.g. the bones of the pelvis could be displayed as semi-transparent, so that even at an oblique angle, one part of the image does not conceal another.

Three Dimensional Image Reconstruction

Because contemporary CT scanners offer isotropic, or near isotropic, resolution, display of images does not need to be restricted to the conventional axial images. Instead, it is possible for a software programme to build a volume by 'stacking' the individual slices one on top of the other. The programme may then display the volume in an alternative manner.

Multiplanar Reconstruction

Multiplanar reconstruction (MPR) is the simplest method of reconstruction. A volume is built by stacking the axial slices. The software then cuts slices through the volume in a different plane (usually orthogonal). Optionally, a special projection method, such as maximum-intensity projection (MIP) or minimum-intensity projection (mIP), can be used to build the reconstructed slices. MPR is frequently used for examining the spine.

Axial images through the spine will only show one vertebral body at a time and cannot reliably show the intervertebral discs. By reformatting the volume, it becomes much easier to visualise the position of one vertebral body in relation to the others.

Modern software allows reconstruction in non-orthogonal (oblique) planes so that the optimal plane can be chosen to display an anatomical structure. This may be particularly useful for visualising the structure of the bronchi as these do not lie orthogonal to the direction of the scan. For vascular imaging, curved-plane reconstruction can be performed.

This allows bends in a vessel to be 'straightened' so that the entire length can be visualised on one image, or a short series of images. Once a vessel has been 'straightened' in this way, quantitative measurements of length and cross sectional area can be made, so that surgery or interventional treatment can be planned.

MIP reconstructions enhance areas of high radiodensity, and so are useful for angiographic studies. mIP reconstructions tend to enhance air spaces so are useful for assessing lung structure.

Chapter 11

Fingerprint Identification

A fingerprint is an impression of the friction ridges of all or any part of the finger. A friction ridge is a raised portion of the epidermis on the palmar (palm and fingers) or plantar (sole and toes) skin, consisting of one or more connected ridge units of friction ridge skin.

These ridges are sometimes known as "dermal ridges" or "dermal papillae". Fingerprints may be deposited in natural secretions from the eccrine glands present in friction ridge skin (secretions consisting primarily of water) or they may be made by ink or other contaminants transferred from the peaks of friction skin ridges to a relatively smooth surface such as a fingerprint card.

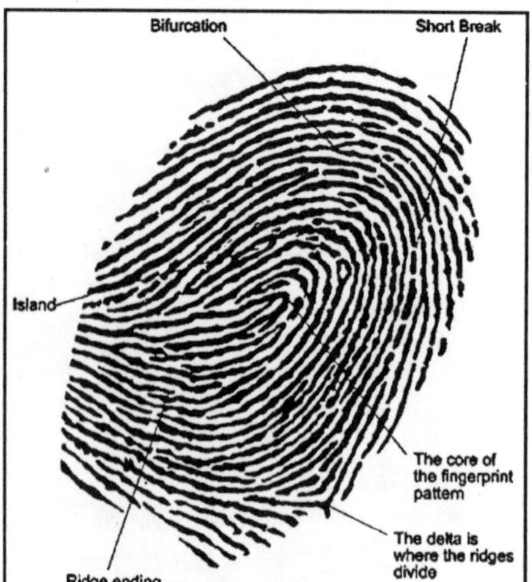

Fig. The Fingerprint

The term fingerprint normally refers to impressions transferred from the pad on the last joint of fingers and thumbs, though fingerprint cards also typically record portions of lower joint areas of the fingers (which are also used to make identifications). On the palmar surface of the hands and feet are raised surfaces called friction ridges. The scientific basis behind friction ridge analysis is the fact that friction ridges are persistent and unique.

Friction ridges are formed during fetal development where their unique characteristics emerge due to genetic and epigenetic factors (maternal diet, pH, temperature, movement of the fetus, etc.). Even identical twins do not have the same fingerprints.

Uniqueness among even identical twins is due to random, or stochastic, effects during fetal development. Stochastic effects have widespread scientific acceptance as a source of uniqueness and have been observed in several animal studies which included fingerprint and other unique traits (hair patterning) between both clones and nuclear transfers.

Friction ridges also persist throughout life in their permanent arrangement barring scarring or injury or until decomposition of the skin following death. Scarring occurs due to damage to the basal layer of the epidermis. Like friction ridges, scars are also persistent throughout life and are re-generated in new layers of skin.

FINGERPRINT IDENTIFICATION

Fingerprint identification (sometimes referred to as *dactyloscopy*) or palmprint identification is the process of comparing questioned and known friction skin ridge impressions from fingers or palms to determine if the impressions are from the same finger or palm. The flexibility of friction ridge skin means that no two finger or palm prints are ever exactly alike (never identical in every detail), even two impressions recorded immediately after each other.

Fingerprint identification occurs when an expert (or an expert computer system operating under threshold scoring rules) determines that two friction ridge impressions originated from the same finger or palm (or toe, sole) to the exclusion of all others.

A known print is the intentional recording of the friction ridges, usually with black printer's ink rolled across a contrasting white background, typically a white card. Friction ridges can also be recorded digitally using a technique called Live-Scan. A latent print is the chance reproduction of the friction ridges deposited on the surface of an item.

Latent prints are often fragmentary and may require chemical methods, powder, or alternative light sources in order to be visualized. When friction ridges come in contact with a surface that is receptive to a print, material on the ridges, such as perspiration, oil, grease, ink, etc. can be transferred to the item. The factors which affect friction ridge impressions are numerous, thereby requiring examiners to undergo extensive and objective study in order to be trained to competency.

Pliability of the skin, deposition pressure, slippage, the matrix, the surface, and the development medium are just some of the various factors which can cause a latent print to appear differently from the known recording of the same friction ridges. Indeed, the conditions of friction ridge deposition are unique and never duplicated. This is another reason as to why extensive and objective study is necessary in order to train examiners to be able to reach competent conclusions.

FINGERPRINT IMAGE

Fingerprint image acquisition is considered as the most critical step of an automated fingerprint authentication system, as it determines the final fingerprint image quality, which has drastic effects on the overall system performance. On the market there are different types of fingerprint readers, but the basic idea behind each capture approach is the measure in some way the physical difference between ridges and valleys. All the proposed methods can be grouped in two major families: solid-state fingerprint readers and optical fingerprint readers.

The procedure to capture a fingerprint using a sensor consists in rolling or touching with the finger onto a sensing area, which according to the used physical principle (capacitive, optical, thermal, etc.) captures the difference between valleys and ridges.

When a finger touches or rolls onto a surface, the elastic skin deforms. The quantity and direction of the pressure applied by the user, the skin conditions and the projection of an irregular 3D object (the finger) onto a 2D flat plane introduce distortions, noise and inconsistencies on the captured fingerprint image.

These problems have been indicated as inconsistent, irreproducible and non-uniform contacts and, during each acquisition, their effect on the same fingerprint results different and uncontrollable.

The representation of the same fingerprint changes every time the finger is placed on the sensor platen, increasing the complexity of the fingerprint matching and representing a negative influence on the system performance with a consequent limited wide-spreading of this biometric technology.

Finger Print Types

Plastic Prints

A plastic print is a friction ridge impression from a finger or palm (or toe/foot) deposited in a material that retains the shape of the ridge detail. Commonly encountered examples are melted candle wax, putty removed from the perimeter of window panes and thick grease deposits on car parts. Such prints are already visible and need no enhancement, but investigators must not overlook the potential that invisible latent prints deposited by accomplices may also be on such surfaces.

After photographically recording such prints, attempts should be made to develop other non-plastic impressions deposited at natural finger/palm secretions (eccrine gland secretions) or contaminates.

Latent Prints

Although the word latent means hidden or invisible, in modern usage for forensic science the term latent prints means any chance or accidental impression left by friction ridge skin on a surface, regardless of whether it is visible or invisible at the time of deposition.

Electronic, chemical and physical processing techniques permit visualization of invisible latent print residue whether they are from natural secretions of the eccrine glands present on friction ridge skin (which produce palmar sweat, sebum, and various kinds of lipids), or whether the impression is in a contaminant such as motor oil, blood, paint, ink, etc. Latent prints may exhibit only a small portion of the surface of the finger and may be smudged, distorted, or both, depending on how they were deposited.

For these reasons, latent prints are an "inevitable source of error in making comparisons," as they generally "contain less clarity, less content, and less undistorted information than a fingerprint taken under controlled conditions, and much, much less detail compared to the actual patterns of ridges and grooves of a finger."

Patent Prints

These are friction ridge impressions of unknown origin which are obvious to the human eye and are caused by a transfer of foreign material on the finger, onto a surface. Because they are already visible they need no enhancement, and are generally photographed instead of being lifted in the same manner as latent prints.

CLASSIFICATION OF FINGERPRINTS

Before computerization replaced manual filing systems in large fingerprint operations, manual fingerprint classification systems were used to categorize fingerprints based on general ridge formations (such as the presence or absence of circular patterns in various fingers).

Thus permitting filing and retrieval of paper records in large collections based on friction ridge patterns independent of name, birth date and other biographic data that persons may misrepresent.

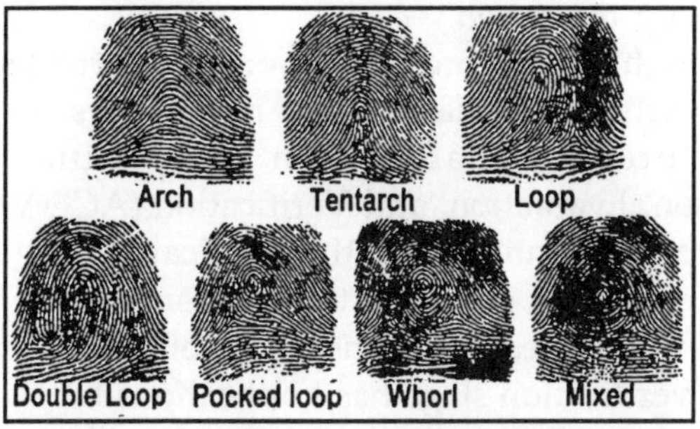

Fig. Classification of Finger Prints

The most popular ten print classification systems include the Roscher system, the Vucetich system, and the Henry Classification System. Of these systems, the Roscher system was developed in Germany and implemented in both Germany and Japan, the Vucetich system was developed in Argentina and implemented throughout South America, and the Henry system was developed in India and implemented in most English-speaking countries..

In the Henry system of classification, there are three basic fingerprint patterns: Arch, Loop and Whorl. There are also more complex classification systems that further break down patterns to plain arches or tented arches. Loops may be radial or ulnar, depending on the side of the hand the tail points towards. Whorls also have sub-group classifications including plain whorls, accidental whorls, double loop whorls, and central pocket loop whorls.

Forensic Fingerprint Evidence

The validity of forensic fingerprint evidence has recently been challenged by academics, judges and the media. While fingerprint identification was an improvement over earlier anthropometric systems, the subjective nature of matching, along with the relatively high error rate of matches when compared to DNA, has made this forensic practice controversial. The words "Reliability" and "Validity" have specific meanings to the scientific community.

Reliability means successive tests bring the same results. Validity means that the results accurately reflect the external criteria being measured.

Although experts are often more comfortable relying on their instincts, this reliance does not always translate into superior predictive ability. In the popular Analysis, Comparison, Evaluation, and Verification (ACE-V) paradigm for fingerprint identification, the verification stage, in which a second examiner confirms the assessment of the original examiner, may increase the consistency of the assessments. But while the verification stage has implications for the reliability of latent print comparisons, it does not assure their validity.

Validity of Forensic Fingerprinting

Despite the absence of objective standards, scientific validation, and adequate statistical studies, a natural question to ask is how well fingerprint examiners actually perform. Proficiency tests do not validate a procedure per se, but they can provide some insight into error rates.

In 1995, the Collaborative Testing Service (CTS) administered a proficiency test that, for the first time, was "designed, assembled, and reviewed" by the International Association for Identification (IAI).The results were disappointing. Four suspect cards with prints of all ten fingers were provided together with seven latents. Of 156 people taking the test, only 68 (44%) correctly classified all seven latents. Overall, the tests contained a total of 48 incorrect identifications.

Errors of this magnitude within a discipline singularly admired and respected for its touted absolute certainty as an identification process have produced chilling and mind-numbing realities. Thirty-four participants, an incredible 22% of those involved, substituted presumed but false certainty for truth. By any measure, this represents a profile of practice that is unacceptable and thus demands positive action by the entire community.

Fingerprints collected at a crime scene, or on items of evidence from a crime, can be used in forensic science to identify suspects, victims and other persons who touched a surface.

Fingerprint identification emerged as an important system within police agencies in the late 19th century, when it replaced anthropometric measurements as a more reliable method for identifying persons having a prior record, often under an alias name, in a criminal record repository. Fingerprint identification was the first forensic discipline (in 1977) to formally institute a professional certification programme for individual experts, including a procedure for decertifying those making errors.

Other forensic disciplines later followed suit in establishing certification programmes whereby certification could be revoked for error. Fingerprint identification effects far more positive identifications of persons worldwide daily than any other human identification procedure.

Some of the discontent over fingerprint evidence may be due to the desire to push the conclusiveness of fingerprint examinations to the same level of certitude as that of DNA analysis. DNA is probability-based inasmuch as an individual is genetically half from the mother's contribution and half from the father's contribution. These genetic contributions are passed down from generation to generation. While pattern type (arch, loops, and whorls) may be inherited, the details of the friction ridges are not.

It cannot be concluded that a person inherited a certain bifurcation from their mother and an ending ridge from their father as the development of these features are completely random. Fingerprints as an analogy of uniqueness has been widely scientifically accepted. Chemists often use the term "fingerprint region" to describe an area of a chemical that can be used to identify it.

Another criticism sometimes leveled at fingerprint practice is that it is a "closed discipline". However, practitioners in the scientific community are generally specialized and may not extend to other areas of science; in this respect, fingerprint scientists are no different from the rest of the scientific community. The fingerprint community asserts that it maintains the need for objectivity and continued research in the area of friction ridge analysis.

Method of Detecting Fingerprints

Since the late nineteenth century, fingerprint identification methods have been used by police agencies around the world to identify both suspected criminals as well as the victims of crime. The basis of the traditional fingerprinting technique is simple. The skin on the palmar surface of the hands and feet forms ridges, so-called papillary ridges, in patterns that are unique to each individual and which do not change over time. Even identical twins do not have identical fingerprints.

Fingerprints on surfaces may be described as patent or latent. Patent fingerprints are left when a substance (such as paint, oil or blood) is transferred from the finger to a surface and are easily photographed without further processing. Latent fingerprints, in contrast, occur when the natural secretions of the skin are deposited on a surface through fingertip contact, and are usually not readily visible. The best way to render latent fingerprints visible, so that they can be photographed, is complex and depends, on the type of surface involved.

It is generally necessary to use a 'developer', usually a powder or chemical reagent, to produce a high degree of visual contrast between the ridge patterns and the surface on which the fingerprint was left. Developing agents depend on the presence of organic deposits for their effectiveness.

However, fingerprints are typically formed by the secretions of the eccrine glands of the fingertips, which principally comprise water and inorganic salts, with only a small proportion of organic material such as urea and amino acids and detecting such fingerprints is far from easy. A further complication is the fact that the organic component of any deposited material is readily destroyed by heat, such as occurs when a gun is fired or a bomb is detonated, when the temperature may reach as high as 500°C.

In contrast, the non-volatile, inorganic component of eccrine secretion remains intact even when exposed to temperatures as high as 600°C. Within the Materials Research Centre, University of Swansea, UK, University of Swansea, UK, Professor Neil McMurray and Dr Geraint Williams have developed a technique that enables fingerprints to be visualised on metallic and electrically conductive surfaces without the need to develop the prints first.

The technique involves the use of an instrument called a scanning Kelvin probe (SKP), which measures the voltage, or electrical potential, at pre-set intervals over the surface of an object on which a fingerprint may have been deposited.

These measurements can then be mapped to produce an image of the fingerprint. A higher resolution image can be obtained by increasing the number of points sampled, but at the expense of the time taken for the process.

A sampling frequency of 20 points per mm is high enough to visualise a fingerprint in sufficient detail for identification purposes and produces a voltage map in 2–3 hours. So far the technique has been shown to work effectively on a wide range of forensically important metal surfaces including iron, steel and aluminium.

While initial experiments were performed on planar, i.e. flat, surfaces, the technique has been further developed to cope with severely non-planar surfaces, such as the warped cylindrical surface of fired cartridge cases. The very latest research from the department has found that physically removing a fingerprint from a metal surface, e.g. by rubbing with a tissue, does not necessarily result in the loss of all fingerprint information.

The reason for this is that the differences in potential that are the basis of the visualisation are caused by the interaction of inorganic salts in the fingerprint deposit and the metal surface and begin to occur as soon as the finger comes into contact with the metal, resulting in the formation of metal–ion complexes that cannot easily be removed. Currently, in crime scene investigations, a decision has to be made at an early stage whether to attempt to retrieve fingerprints through the use of developers or whether to swab surfaces in an attempt to salvage material for DNA fingerprinting.

The two processes are mutually incompatible, as fingerprint developers destroy material that could potentially be used for DNA analysis, and swabbing is likely to make fingerprint identification impossible.

The application of the new SKP fingerprinting technique, which is non-contact and does not require the use of developers, has the potential to allow fingerprints to be retrieved while still leaving intact any material that could subsequently be subjected to DNA analysis.

The University of Swansea group hope to have a forensically usable prototype in the near future and it is intended that eventually the instrument will be manufactured in sufficiently large numbers that it will be widely used by forensic teams on the frontline.

Applications of Finger Print

Electronic fingerprint readers have been introduced for security applications such as identification of computer users (log-in authentication). However, early devices have been discovered to be vulnerable to quite simple methods of deception, such as fake fingerprints cast in gels. In 2006, fingerprint sensors gained popularity in the notebook PC market.

Built-in sensors in ThinkPads, VAIO laptops, and others also double as motion detectors for document scrolling, like the scroll wheel. Another recent use of fingerprints in a day-to-day setting has been the increasing reliance on biometrics in schools where fingerprints and, to a lesser extent, iris scans are used to validate electronic registration, cashless catering, and library access.

This practice is particularly widespread in the UK, where more than 3500 schools currently use such technology, though it is also starting to be adopted in some states in the US.

Footprints

Friction ridge skin present on the soles of the feet and toes (plantar surfaces) is as unique as ridge detail on the fingers and palms (palmar surfaces). When recovered at crime scenes or on items of evidence, sole and toe impressions are used in the same manner as finger and palm prints to effect identifications.

Footprints of infants, along with thumb or index finger prints of mothers, are still commonly recorded in hospitals to assist in verifying the identity of infants. Often, the only identifiable ridge detail in such impressions is from the large toe or adjacent to the large toe, due to the difficulty of recording such fine detail.

When legible ridge detail is lacking, DNA is normally effective (except in instances of chimaerism) for indirectly identifying infants by confirming maternity and paternity of an infant's parents.

It is not uncommon for military records of flight personnel to include bare foot inked impressions. Friction ridge skin protected inside flight boots tends to survive the trauma of a plane crash (and accompanying fire) better than fingers.

Even though the U.S. Armed Forces DNA Identification Laboratory (AFDIL) stores refrigerated DNA samples from all current active duty and reserve personnel, almost all casualty identifications are effected using fingerprints from military ID card records. When friction ridge skin is not available from deceased military personnel, DNA and dental records are used to confirm identity.

Fingerprints in other Species

The Koala is one of the few mammals (other than primates) that have fingerprints. In fact, koala fingerprints are remarkably similar to human fingerprints; even with an electron microscope, it can be quite difficult to distinguish between the two.

Lifestyle Information

The secretions, skin oils and dead cells in the fingerprint contain residues of various chemicals and their metabolites present in the body. These can be detected and used for forensic purposes. The fingerprints of tobacco smokers contain traces of cotinine, a nicotine metabolite; they also contain traces of nicotine itself; however that may be ambiguous as its presence may be caused by mere contact of the finger with a tobacco product.

By treating the fingerprint with gold nanoparticles with attached cotinine antibodies, and then subsequently with fluorescent agent attached to cotinine antibody antibodies, a fingerprint of a smoker becomes fluorescent; non-smoker's fingerprint stays dark.

The same approach is investigated to be used for identifying heavy coffee drinkers, cannabis smokers, and users of various other drugs.

Chapter 12

Heart Attack

Heart Attack, is a medical condition that occurs when the blood supply to a part of the heart is interrupted, most commonly due to rupture of a vulnerable plaque. Other term is Heart Attackis called Acute myocardial infarction (AMI or MI). The resulting ischemia or oxygen shortage causes damage and potential death of heart tissue. It is a medical emergency, and the leading cause of death for both men and women all over the world.

Important risk factors are a previous history of vascular disease such as atherosclerotic coronary heart disease and/or angina, a previous heart attack or stroke, any previous episodes of abnormal heart rhythms or syncope, older age—especially men over 40 and women over 50, smoking, excessive alcohol consumption, the abuse of certain drugs, high triglyceride levels, high LDL ("Low-density lipoprotein") and low HDL ("High density lipoprotein"), diabetes, high blood pressure, obesity, and chronically high levels of stress in certain persons. The term *myocardial infarction* is derived from *myocardium* (the heart muscle) and *infarction* (tissue death due to oxygen starvation).

The phrase "heart attack" is sometimes used incorrectly to describe sudden cardiac death, which may or may not be the result of acute myocardial infarction. A heart attack is different from, but can cause cardiac arrest, which is the stopping of the heartbeat, and cardiac arrhythmia, an abnormal heartbeat.

Classical symptoms of acute myocardial infarction include chest pain (typically radiating to the left arm), shortness of breath, nausea, vomiting, palpitations, sweating, and anxiety. Patients frequently feel suddenly ill. Women often experience different symptoms from men. The most common symptoms of MI in women include shortness of breath, weakness, and fatigue.

Approximately one third of all myocardial infarctions are silent, without chest pain or other symptoms. Immediate treatment for suspected acute myocardial infarction includes oxygen, aspirin, glyceryl trinitrate and pain relief, usually morphine sulfate.

The patient will receive a number of diagnostic tests, such as an electrocardiogram (ECG, EKG), a chest X-ray and blood tests to detect elevated creatine kinase or troponin levels (these are chemical markers released by damaged tissues, especially the myocardium).

Further treatment may include either medications to break down blood clots that block the blood flow to the heart, or mechanically restoring the flow by dilatation or bypass surgery of the blocked coronary artery. Coronary care unit admission allows rapid and safe treatment of complications such as arrhythmia.

MYOCARDIAL INFARCTION

Myocardial infarction is a common presentation of ischemic heart disease. The WHO estimated that in 2002, 12.6 percent of deaths worldwide were from ischemic heart disease. Ischemic heart disease is the leading cause of death in developed countries, but third to AIDS and lower respiratory infections in developing countries.

In India, cardiovascular disease (CVD) is the leading cause of death. The deaths due to CVD in India were 32% of all deaths in 2007 and are expected to rise from 1.17 million in 1990 and 1.59 million in 2000 to 2.03 million in 2010. Although a relatively new epidemic in India, it has quickly become a major health issue with deaths due to CVD expected to double during 1985-2015. Mortality estimates due to CVD vary widely by state, ranging from 10% in Meghalaya to 49% in Punjab (percentage of all deaths). Punjab (49%), Goa (42%), Tamil

Nadu (36%) and Andhra Pradesh (31%) have the highest CVD related mortality estimates.

State-wise differences are correlated with prevalence of specific dietary risk factors in the states. Moderate physical exercise is associated with reduced incidence of CVD in India (those who exercise have less than half the risk of those who don't). CVD also affects Indians at a younger age (in their 30s and 40s) than is typical in other countries.

Risk Factors for Atherosclerosis

Risk factors for atherosclerosis are generally risk factors for myocardial infarction:

- Older age
- Male sex
- Tobacco smoking
- Hypercholesterolemia
- Hyperhomocysteinemia
- Diabetes
- High blood pressure
- Obesity.

Many of these risk factors are modifiable, so many heart attacks can be prevented by maintaining a healthier lifestyle. Physical activity is associated with a lower risk profile. Non-modifiable risk factors include age, sex, and family history of an early heart attack (before the age of 60), which is thought of as reflecting a genetic predisposition.

Socioeconomic factors such as a shorter education and lower income (particularly in women), and living with a partner may also contribute to the risk of MI. To understand epidemiological study results, it's important to note that many factors associated with MI mediate their risk via other factors.

The effect of education is partially based on its effect on income and marital status. Women who use combined oral contraceptive pills have a modestly increased risk of myocardial infarction, especially in the presence of other risk factors, such as smoking. Inflammation is known to be an important step in the process of atherosclerotic plaque formation. C-reactive protein (CRP) is a sensitive but non-specific marker for inflammation.

Elevated CRP blood levels, especially measured with high sensitivity assays, can predict the risk of MI, as well as stroke and development of diabetes. Moreover, some drugs for MI might also reduce CRP levels. The use of high sensitivity CRP assays as a means of screening the general population is advised against, but it may be used optionally at the physician's discretion, in patients who already present with other risk factors or known coronary artery disease.

Whether CRP plays a direct role in atherosclerosis remains uncertain. Inflammation in periodontal disease may be linked coronary heart disease, and since periodontitis is very common, this could have great consequences for public health. Serological studies measuring antibody levels against typical periodontitis-causing bacteria found that such antibodies were more present in subjects with coronary heart disease.

Periodontitis tends to increase blood levels of CRP, fibrinogen and cytokines; thus, periodontitis may mediate its effect on MI risk via other risk factors. Preclinical research suggests that periodontal bacteria can promote aggregation of platelets and promote the formation of foam cells. A role for specific periodontal bacteria has been suggested but remains to be established.

Baldness, hair greying, a diagonal earlobe crease and possibly other skin features are independent risk factors for MI. Their role remains controversial; a common denominator of these signs and the risk of MI is supposed, possibly genetic.

PATHOPHYSIOLOGY

A myocardial infarction occurs when an atherosclerotic plaque slowly builds up in the inner lining of a coronary artery and then suddenly ruptures, totally occluding the artery and preventing blood flow downstream.

Acute myocardial infarction is a type of acute coronary syndrome, which is most frequently (but not always) a manifestation of coronary artery disease. The most common triggering event is the disruption of an atherosclerotic plaque in an epicardial coronary artery, which leads to a clotting cascade, sometimes resulting in total occlusion of the artery. Atherosclerosis is the gradual buildup of cholesterol and fibrous tissue in plaques in the wall of arteries (in this case, the coronary arteries), typically over decades.

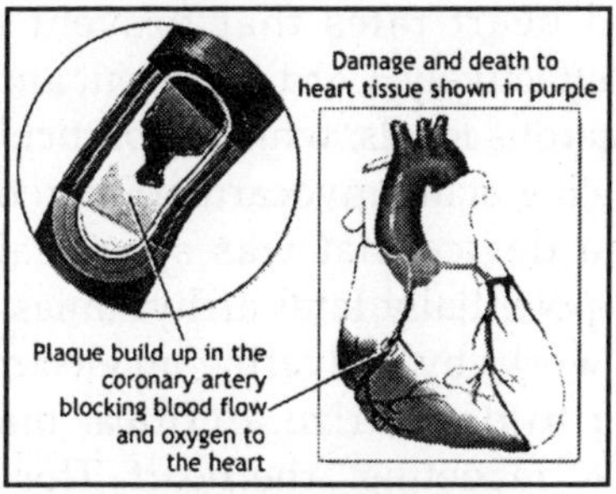

Fig. Myocardial Infarction

Blood stream column irregularities visible on angiographies reflect artery lumen narrowing as a result of decades of advancing atherosclerosis. Plaques can become unstable, rupture, and additionally promote a thrombus (blood clot) that occludes the artery; this can occur in minutes.

When a severe enough plaque rupture occurs in the coronary vasculature, it leads to myocardial infarction (necrosis of downstream myocardium). If impaired blood flow to the heart lasts long enough, it triggers a process called the ischemic cascade; the heart cells die (chiefly through necrosis) and do not grow back.

A collagen scar forms in its place. Recent studies indicate that another form of cell death called apoptosis also plays a role in the process of tissue damage subsequent to myocardial infarction.

As a result, the patient's heart can be permanently damaged. This scar tissue also puts the patient at risk for potentially life threatening arrhythmias. Injured heart tissue conducts electrical impulses more slowly than normal heart tissue. The difference in conduction velocity between injured and uninjured tissue can trigger re-entry or a feedback loop that is believed to be the cause of many lethal arrhythmias. The most serious of these arrhythmias is ventricular fibrillation

(*V-Fib*/VF), an extremely fast and chaotic heart rhythm that is the leading cause of sudden cardiac death.

Another life threatening arrhythmia is ventricular tachycardia (*V-Tach*/VT), which may or may not cause sudden cardiac death. However, ventricular tachycardia usually results in rapid heart rates that prevent the heart from pumping blood effectively. Cardiac output and blood pressure may fall to dangerous levels, which is particularly bad for the patient experiencing acute myocardial infarction. The cardiac defibrillator is a device that was specifically designed to terminate these potentially fatal arrhythmias.

The device works by delivering an electrical shock to the patient in order to depolarize a critical mass of the heart muscle, in effect "rebooting" the heart. This therapy is time dependent, and the odds of successful defibrillation decline rapidly after the onset of cardiopulmonary arrest.

Heart attack rates

Heart attack rates are higher in association with intense exertion, be it psychological stress or physical exertion, especially if the exertion is more intense than the individual usually performs.

Quantitatively, the period of intense exercise and subsequent recovery is associated with about a 6-fold higher myocardial infarction rate (compared with other more relaxed time frames) for people who are physically very fit. For those in poor physical condition, the rate differential is over 35-fold higher.

One observed mechanism for this phenomenon is the increased arterial pulse pressure stretching and relaxation of arteries with each heart beat which, as has been observed with intravascular ultrasound, increases mechanical "shear stress" on atheromas and the likelihood of plaque rupture. Acute severe infection, such as pneumonia, can trigger myocardial infarction. A more controversial link is that between *Chlamydophila pneumoniae* infection and atherosclerosis.

While this intracellular organism has been demonstrated in atherosclerotic plaques, evidence is inconclusive as to whether it can be considered a causative factor. Treatment with antibiotics in patients with proven atherosclerosis has not demonstrated a decreased risk of heart attacks or other coronary vascular diseases.

Myocardial Infarction Symptoms

The onset of symptoms in myocardial infarction (MI) is usually gradual, over several minutes, and rarely instantaneous.

Chest pain is the most common symptom of acute myocardial infarction and is often described as a sensation of tightness, pressure, or squeezing. Chest pain due to ischemia (a lack of blood and hence oxygen supply) of the heart muscle is termed angina pectoris.

Pain radiates most often to the left arm, but may also radiate to the lower jaw, neck, right arm, back, and epigastrium, where it may mimic heartburn. Any group of symptoms compatible with a sudden interruption of the blood flow to the heart are called an acute coronary syndrome. Other conditions such as aortic dissection or pulmonary embolism may present with chest pain and must be considered in the differential diagnosis.

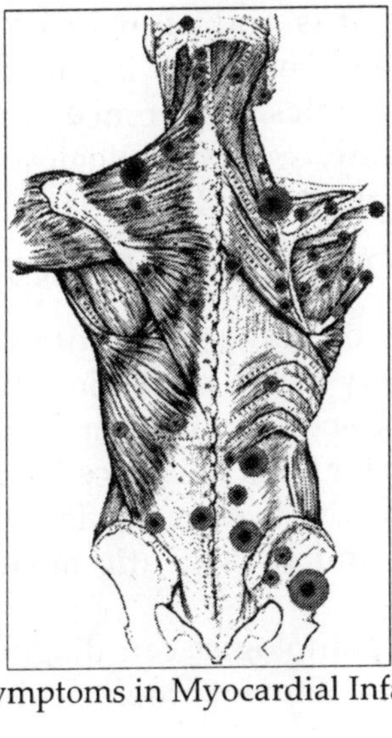

Fig. Symptoms in Myocardial Infarction

Shortness of breath (dyspnea) occurs when the damage to the heart limits the output of the left ventricle, causing left ventricular failure and consequent pulmonary edema. Other symptoms include diaphoresis (an excessive form of sweating), weakness, light-headedness, nausea, vomiting, and palpitations. Loss of consciousness and even sudden death can occur in myocardial infarctions. Women often experience markedly different symptoms than men. The most common symptoms of MI in women include dyspnea, weakness, and fatigue.

Fatigue, sleep disturbances, and dyspnea have been reported as frequently occurring symptoms which may manifest as long as one month before the actual clinically manifested ischemic event. In women, chest pain may be less predictive of coronary ischemia than in men. Approximately half of all MI patients have experienced warning symptoms such as chest pain prior to the infarction. Approximately one fourth of all myocardial infarctions are silent, without chest pain or other symptoms. These cases can be discovered later on electrocardiograms or at autopsy without a prior history of related complaints.

A silent course is more common in the elderly, in patients with diabetes mellitus and after heart transplantation, probably because the donor heart is not connected to nerves of the host. In diabetics, differences in pain threshold, autonomic neuropathy, and psychological factors have been cited as possible explanations for the lack of symptoms.

DIAGNOSIS MYOCARDIAL INFARCTION

The diagnosis of myocardial infarction is made by integrating the history of the presenting illness and physical examination with electrocardiogram findings and cardiac markers (blood tests for heart muscle cell damage). A coronary angiogram allows to visualize narrowings or obstructions on the heart vessels, and therapeutic measures can follow immediately.

At autopsy, a pathologist can diagnose a myocardial infarction based on anatomopathological findings. A chest radiograph and routine blood tests may indicate complications or precipitating causes and are often performed on admittance to an emergency department. New regional wall motion abnormalities on an echocardiogram are also suggestive of a myocardial infarction and are sometimes performed in equivocal cases. Technetium and thallium can be used in nuclear medicine to visualize areas of reduced blood flow and tissue viability, respectively. Technetium is used in a MUGA scan.

Diagnostic Criteria

WHO criteria have classically been used to diagnose MI; a patient is diagnosed with myocardial infarction if two (probable) or three (definite) of the following criteria are satisfied:

- Clinical history of ischaemic type chest pain lasting for more than 20 minutes
- Changes in serial ECG tracings
- Rise and fall of serum cardiac biomarkers such as creatine kinase, troponin I, and lactate dehydrogenase isozymes specific for the heart.

The WHO criteria were refined in 2000 to give more prominence to cardiac biomarkers. According to the new guidelines, a cardiac troponin rise accompanied by either typical symptoms, pathological Q waves, ST elevation or depression or coronary intervention are diagnostic of MI.

Physical Examination

The general appearance of patients may vary according to the experienced symptoms; the patient may be comfortable, or restless and in severe distress with an increased respiratory rate. A cool and pale skin is common and points to vasoconstriction. Some patients have low-grade fever (38–39 °C). Blood pressure may be elevated or decreased, and the pulse can be become irregular. If heart failure ensues, elevated jugular venous pressure and hepatojugular reflux, or swelling of the legs due to peripheral edema may be found on inspection.

Rarely, a cardiac bulge with a pace different from the pulse rhythm can be felt on precordial examination. Various abnormalities can be found on auscultation, such as a third and fourth heart sound, systolic murmurs, paradoxical splitting of the second heart sound, a pericardial friction rub and rales over the lung.

Electrocardiogram

The primary purpose of the electrocardiogram is to detect ischemia or acute coronary injury in broad, symptomatic emergency department populations. However, the standard 12 lead ECG has several limitations. An ECG represents a brief sample in time. Because unstable ischemic syndromes have rapidly changing supply versus demand characteristics, a single ECG may not accurately represent the entire picture.

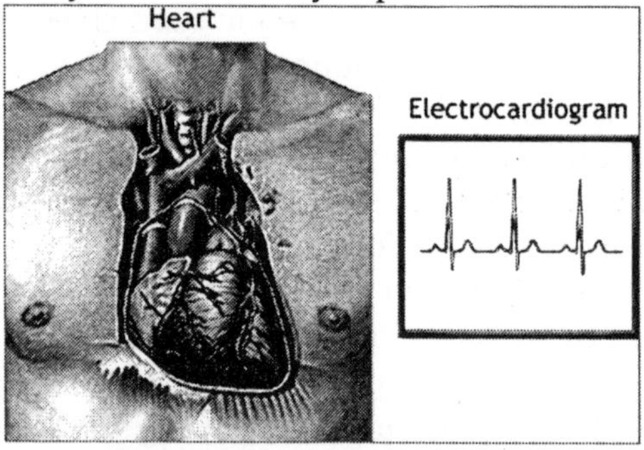

Fig. The Electrocardiogram

It is therefore desirable to obtain serial 12 lead ECGs, particularly if the first ECG is obtained during a pain-free episode. Alternatively, many emergency departments and chest pain centres use computers capable of continuous ST segment monitoring. It should also be appreciated that the standard 12 lead ECG does not directly examine the right ventricle, and does a relatively poor job of examining the posterior basal and lateral walls of the left ventricle.

In particular, acute myocardial infarction in the distribution of the circumflex artery is likely to produce a nondiagnostic ECG. The use of non-standard ECG leads like right-sided lead V4R and posterior leads V7, V8, and V9 may improve sensitivity for right ventricular and posterior myocardial infarction.

In spite of these limitations, the 12 lead ECG stands at the centre of risk stratification for the patient with suspected acute myocardial infarction. Mistakes in interpretation are relatively common, and the failure to identify high risk features has a negative effect on the quality of patient care. The 12 lead ECG is used to classify patients into one of three groups:

- Those with ST segment elevation or new bundle branch block (suspicious for acute injury and a possible candidate for acute reperfusion therapy with thrombolytics or primary PCI),
- Those with ST segment depression or T wave inversion (suspicious for ischemia), and
- Those with a so-called non-diagnostic or normal ECG.

A normal ECG does not rule out acute myocardial infarction. Sometimes the earliest presentation of acute myocardial infarction is the hyperacute T wave, which is treated the same as ST segment elevation. In practice this is rarely seen, because it only exists for 2-30 minutes after the onset of infarction. Hyperacute T waves need to be distinguished from the peaked T waves associated with hyperkalemia.

The current guidelines for the ECG diagnosis of acute myocardial infarction require at least 1 mm (0.1 mV) of ST segment elevation in 2 or more anatomically contiguous leads. This criterion is problematic, however, as acute myocardial infarction is not the most common cause of ST segment elevation in chest pain patients. In addition, over 90% of healthy men have at least 1 mm (0.1 mV) of ST segment elevation in at least one precordial lead.

The clinician must therefore be well versed in recognizing the so-called ECG mimics of acute myocardial infarction, which include left ventricular hypertrophy, left bundle branch block, paced rhythm, benign early repolarization, pericarditis, hyperkalemia, and ventricular aneurysm. Left bundle branch block and pacing can interfere with the electrocardiographic diagnosis of acute myocadial infarction.

The GUSTO investigators Sgarbossa et al. developed a set of criteria for identifying acute myocardial infarction in the presence of left bundle branch block and paced rhythm. They include concordant ST segment elevation > 1 mm (0.1 mV), discordant ST segment elevation > 5 mm (0.5 mV), and concordant ST segment depression in the left precordial leads. The presence of reciprocal changes on the 12 lead ECG may help distinguish true acute myocardial infarction from the mimics of acute myocardial infarction.

The contour of the ST segment may also be helpful, with a straight or upwardly convex (non-concave) ST segment favouring the diagnosis of acute myocardial infarction. The constellation of leads with ST segment elevation enables the clinician to identify what area of the heart is injured, which in turn helps predict the so-called culprit artery. As the myocardial infarction evolves, there may be loss of R wave height and development of pathological Q waves.

T wave inversion may persist for months or even permanently following acute myocardial infarction. Typically, however, the T wave recovers, leaving a pathological Q wave as the only remaining evidence that an acute myocardial infarction has occurred.

Cardiac Enzymes

Cardiac markers or cardiac enzymes are proteins from cardiac tissue found in the blood. These proteins are released into the bloodstream when damage to the heart occurs, as in the case of a myocardial infarction.

Until the 1980s, the enzymes SGOT and LDH were used to assess cardiac injury. Then it was found that disproportional elevation of the *MB* subtype of the enzyme creatine kinase (CK) was very specific for myocardial injury.

Current guidelines are generally in favour of troponin sub-units I or T, which are very specific for the heart muscle and are thought to rise before permanent injury develops.

Elevated troponins in the setting of chest pain may accurately predict a high likelihood of a myocardial infarction in the near future.

The diagnosis of myocardial infarction requires two out of three components. When damage to the heart occurs, levels of cardiac markers rise over time, which is why blood tests for them are taken over a 24 hour period.

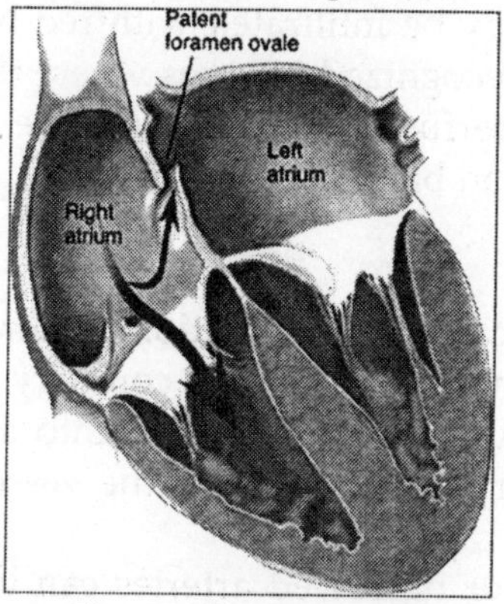

Fig. Cardiac Enzymes

Because these enzyme levels are not elevated immediately following a heart attack, patients presenting with chest pain are generally treated with the assumption that a myocardial infarction has occurred and then evaluated for a more precise diagnosis.

Histopathology

Histopathological examination of the heart may reveal infarction at autopsy. Under the microscope, myocardial infarction presents as a circumscribed area of ischemic, coagulative necrosis (cell death). On gross examination, the infarct is not identifiable within the first 12 hours. Although earlier changes can be discerned using electron microscopy, one of the earliest changes under a normal microscope are so-called *wavy fibres*.

Subsequently, the myocyte cytoplasm becomes more eosinophilic (pink) and the cells lose their transversal striations, with typical changes and eventually loss of the cell nucleus. The interstitium at the margin of the infarcted area is initially infiltrated with neutrophils, then with lymphocytes and macrophages, who phagocytose ("eat") the myocyte debris.

The necrotic area is surrounded and progressively invaded by granulation tissue, which will replace the infarct with a fibrous (collagenous) scar (which are typical steps in wound healing).

The interstitial space (the space between cells outside of blood vessels) may be infiltrated with red blood cells. These features can be recognized in cases where the perfusion was not restored; reperfused infarcts can have other hallmarks, such as contraction band necrosis.

Angiography

In difficult cases or in situations where intervention to restore blood flow is appropriate, coronary angiography can be performed. A catheter is inserted into an artery (usually the femoral artery) and pushed to the vessels supplying the heart.

Obstructed or narrowed arteries can be identified, and angioplasty applied as a therapeutic measure. Angioplasty requires extensive skill, especially in emergency settings, and may not always be available out of hours. It is commonly performed by interventional cardiologists.

Automatic External Defibrillation (AED)

Since the publication of data showing that the availability of automated external defibrillators (AEDs) in public places may significantly increase chances of survival, many of these have been installed in public buildings, public transport facilities, and in non-ambulance emergency vehicles (e.g. police cars and fire engines). AEDs analyse the heart's rhythm and determine whether the rhythm is amenable to defibrillation ("shockable"), as in ventricular tachycardia and ventricular fibrillation.

Emergency Services

Emergency Medical Services (EMS) Systems vary considerably in their ability to evaluate and treat patients with suspected acute myocardial infarction. Some provide as little as first aid and early defibrillation. Others employ highly trained paramedics with sophisticated technology and advanced protocols. Early access to EMS is promoted by a 9-1-1 system currently available to 90% of the population in the United States.

Most are capable of providing oxygen, IV access, sublingual nitroglycerine, morphine, and aspirin. Some are capable of providing thrombolytic therapy in the prehospital setting.

With primary PCI emerging as the preferred therapy for ST segment elevation myocardial infarction, EMS can play a key role in reducing door to balloon intervals (the time from presentation to a hospital ER to the restoration of coronary artery blood flow) by performing a 12 lead ECG in the field and using this information to triage the patient to the most appropriate medical facility. In addition, the 12 lead ECG can be transmitted to the receiving hospital, which enables time saving decisions to be made prior to the patient's arrival.

This may include a "cardiac alert" or "STEMI alert" that calls in off duty personnel in areas where the cardiac cath lab is not staffed 24 hours a day. Even in the absence of a formal alerting programme, prehospital 12 lead ECGs are independently associated with reduced door to treatment intervals in the emergency department.

Wilderness First Aid

In wilderness first aid, a possible heart attack justifies evacuation by the fastest available means, including MEDEVAC, even in the earliest or precursor stages. The patient will rapidly be incapable of further exertion and have to be carried out.

Air Travel

Certified personnel traveling by commercial aircraft may be able to assist an MI patient by using the on-board first aid kit, which may contain some cardiac drugs (such as glyceryl trinitrate spray, aspirin, or opioid painkillers) and oxygen. Pilots may divert the flight to land at a nearby airport. Cardiac monitors are being introduced by some airlines, and they can be used by both on-board and ground-based physicians.

TREATMENT

A heart attack is a medical emergency which demands both immediate attention and activation of the emergency medical services. The ultimate goal of the management in the acute phase of the disease is to salvage as much myocardium as possible and prevent further complications. As time passes, the risk of damage to the heart muscle increases; hence the phrase that in myocardial infarction, "time is muscle," and time wasted is muscle lost.

The treatments itself may have complications. If attempts to restore the blood flow are initiated after a critical period of only a few hours, the result is reperfusion injury instead of amelioration. Other treatment modalities may also cause complications; the use of antithrombotics for example carries an increased risk of bleeding.

First Line

Oxygen, aspirin, glyceryl trinitrate (nitroglycerin) and analgesia (usually morphine, hence the popular mnemonic *MONA, morphine, oxygen, nitro, aspirin*) are administered as soon as possible. In many areas, first responders can be trained to administer these prior to arrival at the hospital.

Morphine is the preferred pain relief drug due to its ability to dilate blood vessels, which aids in blood flow to the heart as well as its pain relief properties. Of the first line agents, only aspirin has been proven to decrease mortality. Once the diagnosis of myocardial infarction is confirmed, other pharmacologic agents are often given.

These include beta blockers, anticoagulation (typically with heparin), and possibly additional antiplatelet agents such as clopidogrel.

These agents are typically not started until the patient is evaluated by an emergency room physician or under the direction of a cardiologist. These agents can be used regardless of the reperfusion strategy that is to be employed. While these agents can decrease mortality in the setting of an acute myocardial infarction, they can lead to complications and potentially death if used in the wrong setting.

Reperfusion

The concept of reperfusion has become so central to the modern treatment of acute myocardial infarction, that we are said to be in the reperfusion era. Patients who present with suspected acute myocardial infarction and ST segment elevation (STEMI) or new bundle branch block on the 12 lead ECG are presumed to have an occlusive thrombosis in an epicardial coronary artery.

They are therefore candidates for immediate reperfusion, either with thrombolytic therapy, percutaneous coronary intervention (PCI) or when these therapies are unsuccessful, bypass surgery. Individuals without ST segment elevation are presumed to be experiencing either unstable angina (UA) or non-ST segment elevation myocardial infarction (NSTEMI). They receive many of the same initial therapies and are often stabilized with antiplatelet drugs and anticoagulated.

If their condition remains (hemodynamically) stable, they can be offered either late coronary angiography with subsequent restoration of blood flow (revascularization), or non-invasive stress testing to determine if there is significant ischemia that would benefit from revascularization. If hemodynamic instability develops in individuals with NSTEMIs, they may undergo urgent coronary angiography and subsequent revascularization.

The use of thrombolytic agents is contraindicated in this patient subset, however. The basis for this distinction in treatment regimens is that ST segment elevations on an ECG are typically due to complete occlusion of a coronary artery. On the other hand, in NSTEMIs there is typically a sudden narrowing of a coronary artery with preserved (but diminished) flow to the distal myocardium.

Anticoagulation and antiplatelet agents are given to prevent the narrowed artery from occluding. At least 10% of patients with STEMI don't develop myocardial necrosis (as evidenced by a rise in cardiac markers) and subsequent q waves on EKG after reperfusion therapy. Such a successful restoration of flow to the infarct-related artery during an acute myocardial infarction is known as "aborting" the myocardial infarction. If treated within the hour, about 25% of STEMIs can be aborted.

Thrombolytic Therapy

Thrombolytic therapy is indicated for the treatment of STEMI if the drug can be administered within 12 hours of the onset of symptoms, the patient is eligible based on exclusion criteria, and primary PCI is not immediately available. The effectiveness of thrombolytic therapy is highest in the first 2 hours. After 12 hours, the risk associated with thrombolytic therapy outweighs any benefit.

Because irreversible injury occurs within 2–4 hours of the infarction, there is a limited window of time available for reperfusion to work. Thrombolytic drugs are contraindicated for the treatment of unstable angina and NSTEMI and for the treatment of individuals with evidence of cardiogenic shock.

Although no perfect thrombolytic agent exists, an ideal thrombolytic drug would lead to rapid reperfusion, have a high sustained patency rate, be specific for recent thrombi, be easily and rapidly administered, create a low risk for intra-cerebral and systemic bleeding, have no antigenicity, adverse hemodynamic effects, or clinically significant drug interactions, and be cost effective.

Currently available thrombolytic agents include streptokinase, urokinase, and alteplase (recombinant tissue plasminogen activator, rtPA). More recently, thrombolytic agents similar in structure to rtPA such as reteplase and tenecteplase have been used. These newer agents boast efficacy at least as good as rtPA with significantly easier administration. The thrombolytic agent used in a particular individual is based on institution preference and the age of the patient. Depending on the thrombolytic agent being used, adjuvant anticoagulation with heparin or low molecular weight heparin may be of benefit.

With tPA and related agents (reteplase and tenecteplase), heparin is needed to maintain coronary artery patency. Because of the anticoagulant effect of fibrinogen depletion with streptokinase and urokinase treatment, it is less necessary there. Intracranial bleeding (ICB) and subsequent cerebrovascular accident (CVA) is a serious side effect of thrombolytic use. The risk of ICB is dependent on a number of factors, including a previous episode of intracranial bleed, age of the individual, and the thrombolytic regimen that is being used.

In general, the risk of ICB due to thrombolytic use for the treatment of an acute myocardial infarction is between 0.5 and 1 percent. Thrombolytic therapy to abort a myocardial infarction is not always effective. The degree of effectiveness of a thrombolytic agent is dependent on the time since the myocardial infarction began, with the best results occurring if the thrombolytic agent is used within two hours of the onset of symptoms. If the individual presents more than 12 hours after symptoms commenced, the risk of intracranial bleed are considered higher than the benefits of the thrombolytic agent.

Failure rates of thrombolytics can be as high as 20% or higher. In cases of failure of the thrombolytic agent to open the infarct-related coronary artery, the patient is then either treated conservatively with anticoagulants and allowed to "complete the infarction" or percutaneous coronary intervention is then performed. Percutaneous coronary intervention in this setting is known as "rescue PCI" or "salvage PCI". Complications, particularly bleeding, are significantly higher with rescue PCI than with primary PCI due to the action of the thrombolytic agent.

Coronary Intervention

The benefit of prompt, expertly performed primary percutaneous coronary intervention over thrombolytic therapy for acute ST elevation myocardial infarction is now well established. Logistic and economic obstacles seem to hinder a more widespread application of percutaneous coronary intervention (PCI) via cardiac catheterization, although the feasibility of regionalized PCI for STEMI is currently being explored in the United States.

The use of percutaneous coronary intervention as a therapy to abort a myocardial infarction is known as primary

PCI. The goal of primary PCI is to open the artery as soon as possible, and preferably within 90 minutes of the patient presenting to the emergency room. This time is referred to as the door-to-balloon time. The current guidelines in the United States restrict primary PCI to hospitals with available emergency bypass surgery as a backup, but this is not the case in other parts of the world. Primary PCI involves performing a coronary angiogram to determine the anatomical location of the infarcting vessel, followed by balloon angioplasty (and frequently deployment of an intracoronary stent) of the thrombosed arterial segment.

In some settings, an extraction catheter may be used to attempt to aspirate (remove) the thrombus prior to balloon angioplasty. While the use of intracoronary stents do not improve the short term outcomes in primary PCI, the use of stents is widespread because of the decreased rates of procedures to treat restenosis compared to balloon angioplasty. Adjuvant therapy during primary PCI include intravenous heparin, aspirin, and clopidogrel.

The use of glycoprotein IIb/IIIa inhibitors are often used in the setting of primary PCI to reduce the risk of ischemic complications during the procedure. Due to the number of antiplatelet agents and anticoagulants used during primary PCI, the risk of bleeding associated with the procedure are higher than during an elective PCI.

Coronary Artery Bypass Surgery

Coronary artery bypass surgery during mobilization (freeing) of the right coronary artery from its surrounding tissue, adipose tissue (yellow). The tube visible at the bottom is the aortic cannula (returns blood from the HLM). The tube above it (obscured by the surgeon on the right) is the venous cannula (receives blood from the body). The patient's heart is stopped and the aorta is cross-clamped. The patient's head is at the bottom. Despite the guidelines, emergency bypass surgery for the treatment of an acute myocardial infarction (MI) is less common then PCI or medical management. In an analysis of patients in the U.S.

National Registry of Myocardial Infarction (NRMI) from January 1995 to May 2004, the percentage of patients with cardiogenic shock treated with primary PCI rose from 27.4% to 54.4%, while the increase in CABG treatment was only from 2.1% to 3.2%. Emergency coronary artery bypass graft surgery (CABG) is usually undertaken to simultaneously treat a mechanical complication, such as a ruptured papillary muscle, or a ventricular septal defect, with ensueing cardiogenic shock. In uncomplicated MI, the mortality rate can be high when the surgery is performed immediately following the infarction.

If this option is entertained, the patient should be stabilized prior to surgery, with supportive interventions such as the use of an intra-aortic balloon pump. In patients developing cardiogenic shock after a myocardial infarction, both PCI and CABG are satisfactory treatment options, with similar survival rates.

Coronary artery bypass surgery involves an artery or vein from the patient being implanted to bypass narrowings or occlusions on the coronary arteries. Several arteries and veins can be used, however internal mammary artery grafts have demonstrated significantly better long-term patency rates than great saphenous vein grafts.

In patients with two or more coronary arteries affected, bypass surgery is associated with higher long-term survival rates compared to percutaneous interventions. In patients with single vessel disease, surgery is comparably safe and effective, and may be a treatment option in selected cases. Bypass surgery has higher costs initially, but becomes cost-effective in the long term. A surgical bypass graft is more invasive initially but bears less risk of recurrent procedures (but these may be again minimally invasive).

Monitoring for Arrhythmias

Additional objectives are to prevent life-threatening arrhythmias or conduction disturbances. This requires monitoring in a coronary care unit and protocolised administration of antiarrhythmic agents. Antiarrhythmic agents are typically only given to individuals with life-threatening arrhythmias after a myocardial infarction and not to suppress the ventricular ectopy that is often seen after a myocardial infarction.

Rehabilitation

Cardiac rehabilitation aims to optimize function and quality of life in those afflicted with a heart disease. This can be with the help of a physician, or in the form of a cardiac rehabilitation programme.

Physical exercise is an important part of rehabilitation after a myocardial infarction, with beneficial effects on cholesterol levels, blood pressure, weight, stress and mood. Some patients become afraid of exercising because it might trigger another infarct.

Patients are stimulated to exercise, and should only avoid certain exerting activities such as shovelling. Local authorities may place limitations on driving motorised vehicles. Some people are afraid to have sex after a heart attack.

Most people can resume sexual activities after 3 to 4 weeks. The amount of activity needs to be dosed to the patients possibilities.

SECONDARY PREVENTION

The risk of a recurrent myocardial infarction decreases with strict blood pressure management and lifestyle changes, chiefly smoking cessation, regular exercise, a sensible diet for patients with heart disease, and limitation of alcohol intake. Patients are usually commenced on several long-term medications post-MI, with the aim of preventing secondary cardiovascular events such as further myocardial infarctions, congestive heart failure or cerebrovascular accident (CVA). Unless contraindicated, such medications may include:

- Antiplatelet drug therapy such as aspirin and/or clopidogrel should be continued to reduce the risk of plaque rupture and recurrent myocardial infarction. Aspirin is first-line, owing to its low cost and comparable efficacy, with clopidogrel reserved for patients intolerant of aspirin. The combination of clopidogrel and aspirin may further reduce risk of cardiovascular events, however the risk of hemorrhage is increased.

- Beta blocker therapy such as metoprolol or carvedilol should be commenced. These have been particularly beneficial in high-risk patients such as those with left ventricular dysfunction and/or continuing cardiac ischaemia. â-Blockers decrease mortality and morbidity. They also improve symptoms of cardiac ischemia in NSTEMI.
- ACE inhibitor therapy should be commenced 24–48 hours post-MI in hemodynamically-stable patients, particularly in patients with a history of MI, diabetes mellitus, hypertension, anterior location of infarct, and/or evidence of left ventricular dysfunction. ACE inhibitors reduce mortality, the development of heart failure, and decrease ventricular remodelling post-MI.
- Statin therapy has been shown to reduce mortality and morbidity post-MI. The effects of statins may be more than their LDL lowering effects. The general consensus is that statins have plaque stabilization and multiple other ("pleiotropic") effects that may prevent myocardial infarction in addition to their effects on blood lipids.
- The aldosterone antagonist agent eplerenone has been shown to further reduce risk of cardiovascular death post-MI in patients with heart failure and left ventricular dysfunction, when used in conjunction with standard therapies above.
- Omega-3 fatty acids, commonly found in fish, have been shown to reduce mortality post-MI. While the mechanism by which these fatty acids decrease mortality is unknown, it has been postulated that the survival benefit is due to electrical stabilization and the prevention of ventricular fibrillation. However, further studies in a high-risk subset have not shown a clear-cut decrease in potentially fatal arrhythmias due to omega-3 fatty acids.

New Therapies under Investigation

Patients who receive stem cell treatment by coronary artery injections of stem cells derived from their own bone marrow after a myocardial infarction (MI) show improvements in left ventricular ejection fraction and end-diastolic volume not seen with placebo. The larger the initial infarct size, the greater the effect of the infusion.

Clinical trials of progenitor cell infusion as a treatment approach to ST elevation MI are proceeding. There are currently 3 biomaterial and tissue engineering approaches for the treatment of MI, but these are in an even earlier stage of medical research, so many questions and issues need to be addressed before they can be applied to patients. The first involves polymeric left ventricular restraints in the prevention of heart failure.

The second utilizes *in vitro* engineered cardiac tissue, which is subsequently implanted *in vivo*. The final approach entails injecting cells and/or a scaffold into the myocardium to create *in situ* engineered cardiac tissue.

COMPLICATIONS

Complications may occur immediately following the heart attack (in the acute phase), or may need time to develop (a chronic problem). After an infarction, an obvious complication is a second infarction, which may occur in the domain of another atherosclerotic coronary artery, or in the same zone if there are any live cells left in the infarct.

Congestive Heart Failure

A myocardial infarction may compromise the function of the heart as a pump for the circulation, a state called heart failure.

There are different types of heart failure; left- or right-sided (or bilateral) heart failure may occur depending on the affected part of the heart, and it is a low-output type of failure.

If one of the heart valves is affected, this may cause dysfunction, such as mitral regurgitation in the case of left-sided MI. The incidence of heart failure is particularly high in patients with diabetes and requires special management strategies.

Myocardial Rupture

Myocardial rupture is most common three to five days after myocardial infarction, commonly of small degree, but may occur one day to three weeks later. In the modern era of early revascularization and intensive pharmacotherapy as treatment for MI, the incidence of myocardial rupture is about 1% of all MIs. This may occur in the free walls of the ventricles, the septum between them, the papillary muscles, or less commonly the atria.

Rupture occurs because of increased pressure against the weakened walls of the heart chambers due to heart muscle that cannot pump blood out effectively. The weakness may also lead to ventricular aneurysm, a localized dilation or ballooning of the heart chamber. Risk factors for myocardial rupture include completion of infarction (no revascularization performed), female sex, advanced age, and a lack of a previous history of myocardial infarction.

In addition, the risk of rupture is higher in individuals who are revascularized with a thrombolytic agent than with PCI. The shear stress between the infarcted segment and the surrounding normal myocardium (which may be hypercontractile in the post-infarction period) makes it a nidus for rupture.

Rupture is usually a catastrophic event that may result a life-threatening process known as cardiac tamponade, in which blood accumulates within the pericardium or heart sac, and compresses the heart to the point where it cannot pump effectively.

Rupture of the intraventricular septum (the muscle separating the left and right ventricles) causes a ventricular septal defect with shunting of blood through the defect from the left side of the heart to the right side of the heart. Rupture of the papillary muscle may also lead to acute mitral regurgitation and subsequent pulmonary edema and possibly even cardiogenic shock.

Life-threatening Arrhythmia

Since the electrical characteristics of the infarcted tissue change, arrhythmias are a frequent complication. The re-entry phenomenon may cause rapid heart rates (ventricular tachycardia and even ventricular fibrillation), and ischemia in the electrical conduction system of the heart may cause a complete heart block (when the impulse from the sinoatrial node, the normal cardiac pacemaker, does not reach the heart chambers).

Pericarditis

As a reaction to the damage of the heart muscle, inflammatory cells are attracted. The inflammation may reach out and affect the heart sac. This is called pericarditis. In Dressler's syndrome, this occurs several weeks after the initial event.

Cardiogenic Shock

A complication that may occur in the acute setting soon after a myocardial infarction or in the weeks following it is cardiogenic shock.

Cardiogenic shock is defined as a hemodynamic state in which the heart cannot produce enough of a cardiac output to supply an adequate amount of oxygenated blood to the tissues of the body.

While the data on performing interventions on individuals with cardiogenic shock is sparse, trial data suggests a long-term mortality benefit in undergoing revascularization if the individual is less than 75 years old and if the onset of the acute myocardial infarction is less than 36 hours and the onset of cardiogenic shock is less than 18 hours.

If the patient with cardiogenic shock is not going to be revascularized, aggressive hemodynamic support is warranted, with insertion of an intra-aortic balloon pump if not contraindicated.

If diagnostic coronary angiography does not reveal a culprit blockage that is the cause of the cardiogenic shock, the prognosis is poor.

PROGNOSIS

The prognosis for patients with myocardial infarction varies greatly, depending on the patient, the condition itself and the given treatment. Using simple variables which are immediately available in the emergency room, patients with a higher risk of adverse outcome can be identified.

One study found that 0.4% of patients with a low risk profile had died after 90 days, whereas the mortality rate in high risk patients was 21.1%. Although studies differ in the identified variables, some of the more reproduced risk stratifiers include age, hemodynamic parameters (such as heart failure, cardiac arrest on admission, systolic blood pressure, or Killip class of two or greater), ST-segment deviation, diabetes, serum creatinine concentration, peripheral vascular disease and elevation of cardiac markers.

Assessment of left ventricular ejection fraction may increase the predictive power of some risk stratification models. The prognostic importance of Q-waves is debated. Prognosis is significantly worsened if a mechanical complication (papillary muscle rupture, myocardial free wall rupture, and so on) were to occur. There is evidence that case fatality of myocardial infarction has been improving over the years in all ethnicities.

LEGAL IMPLICATIONS

A myocardial infarction is generally a disease, but may sometimes be an injury. This has implications for no-fault insurance schemes such as workers' compensation. A heart attack is generally not covered; however, it may be a work-related injury if it results, from unusual emotional stress or unusual exertion. Additionally, in some jurisdictions, heart attacks suffered by persons in particular occupations such as police officers may be classified as line-of-duty injuries by statute or policy.

The last fifty years have completely changed the way biological and medical researchers can study and understand life, its development from conception to death, susceptibility to infectious and inherited diseases, in short, the molecular mechanisms of metabolic processes. One reason that brought about this understanding lies in the ability to access the information contained in biological macromolecules. Information stored in the structure of molecules is a function of their physical and chemical properties.

A second and more important reason is the ability to manipulate this information by virtue of changing the structures of macromolecules - proteins, nucleic acids, or polysaccharides.

The advancement of molecular biology has been the driving force behind these changes in the biomedical sciences. But the functional manipulation of biological material could not generate much of what is done today by the pharmaceutical industry, if it were not for the preceding developments in physics and chemistry during the 19th and 20th centuries including thermodynamics, statistical mechanics, and the nature of the chemical bond.

The reductionist's approach - the study of chemistry and physics of life - created an enormous wealth of biochemical and genetic data available for the rational design of drugs and the manipulation of the genome. While DNA is the storage of hereditary information, proteins and RNA are its agents, accessing and executing the genetic programmes. The mechanism of protein function is simple; proteins accelerate chemical reactions (as enzymes) by providing optimal binding to substrates, or drastically improve solubility and target specific binding (as receptors) of small ligand molecules.

All catalytic activity and ligand binding occurs on solvent accessible protein surfaces provided by preferential molecular interactions. These molecular interactions are electrostatic in nature. The strength of these interactions, the forces among atoms, can be categorized according to their thermodynamic and kinetic behaviour and is defined as affinity. The conformational precision of interaction leads to selective binding and is defined as specificity. Both properties are directly dependent on the physico-chemical properties of the solvent of life - water.

Chapter 13

Principles of Molecular Recognition

The function of most proteins is controlled by small molecule ligands that reversibly bind to proteins and either stimulate or inhibit their activity. Because different areas of research have studied different kinds of proteins, more than one nomenclature for these small ligand molecules are used. For enzymes which catalyze chemical reactions, natural ligands are the substrates which have to bind before they are being chemically processed into products.

Table. Classification of Ligands

Target structure	Positive effector	Negative effector
Enzyme	Substrate	Competitive inhibitor
Receptor	Agonist	Antagonist

Catalytic reactions can be suppressed by competitive inhibitors which bind to the same location on the surface of an enzyme as the substrate. This location is called the active site. Receptors (e.g. cell surface proteins, nuclear DNA-binding proteins) bind ligands without chemically modifying them. Instead binding induces a conformational change in the receptor protein that can trigger a chemical reaction of a substrate bound somewhere else on the same protein or affect the binding affinity of a second molecule that interacts with the receptor.

This process is known as allosteric regulation. Ligands activating receptors are called agonists, while competitive inhibitors of these ligands are called antagonists. Binding events are characterized as reversible chemical equilibrium and binding (and thus effect) of both agonists and antagonists are concentration dependent.

The affinity of the ligand for the binding site can be quantified by the equilibrium constant of binding (association constant KA) or unbinding (dissociation constant KD) over an effective concentration range, i.e., where an agonist induces an effect, or an antagonist can block agonist activity.

Affinity tells us how strong a ligand binds to its receptor, is related to the Gibbs free energy of binding. Affinity is a macroscopic property if binding representing an averaged behaviour of a very large number of events that are the result of an often complex series of events and molecular interactions.

The latter are microscopic properties of molecular structures and are described by non-covalent bond structures. Bridging the qualitative difference between macroscopic properties (thermodynamic quantities and kinetic data) and microscopic structural information (chemical bonds, electron density maps) is the biggest obstacle to predicting functional aspects of ligand-receptor interaction.

Molecular similarity Space

The following discussion refers to the structure of a stable complex between a ligand and its receptor which is a microscopic description.

The strength of an interaction (its affinity, which is a macroscopic description) depends on the complementarity of the physico-chemical properties of atoms that bind, i.e., protein surface and ligand structure.

Excluding catalytic mechanisms from the discussion, two classes of molecular properties important for binding can readily be distinguished:

- Shape or volume
- Surface potential

Talking about these properties, chemists refer to them as the molecular similarity space which can now more precisely be described as:

Atom pair matching function	(shape or volume)	weak interactions
Charge matching function	(surface potential)	strong interactions

It is intuitive to think that simple binding has to do with similarity (or complementarity) of properties and structures such that the higher a similarity the more specific the recognition will be. Atom pair matching function mostly refers to Van der Waals surfaces of molecules and hydrogen bonds. Both interactions are weak and are effective only over very short distances in terms of potential energy function, i.e., they are short-range interactions that can be easily broken, but help define the conformational specificity of an interaction.

This is particularly true for the hydrogen bond, which has a directional quality related to its strength of interaction. Recognition by a ligand of its receptor binding site can be envisaged as a result of orientational and translational movement of the ligand within the electric surface potential field of the receptor. A specific interaction is encountered when the orientation of the ligand fits complementary physical properties on the receptor surface. A combination of any of these four physical properties summarized in the table above defines multivariate surfaces.

This can obviously lead to complex surface structures or binding motifs, specially for large contact surfaces such as found between proteins, where one protein is the 'ligand' and the other the 'receptor'. Protein-protein interaction is relevant for any enzyme or receptor complex, cytoskeletal structures, or chromatin structures. In the present context, peptide ligands are among those agonists providing the largest variability in similarity space. Since protein surfaces are determined by their amino acids residues, binding surfaces can be mathematically described as sequence space.

An artificial peptide sequence space for the development of novel antibiotics has recently been achieved using cyclic peptides that self-assemble into peptide nanotubes. It is the complementarity of these motifs between receptor and ligand that determines the specificity of the interaction. The electrostatic force between ligand and receptor helps define the affinity of the interaction. There are long range and short range interactions.

Hydrophobic interactions and hydrogen bonds are short range interactions based on induced-dipole and molecular dipole moments, respectively. Electrostatic interactions are long range meaning that electric fields can be sensed several angstroms away from the point charge. The strength (and effective distance) of these interactions is a function of the dielectric property of the environment. Water molecules are able to shield locale charges and dipoles reducing the range of their electric field forces.

In a hydrophobic environment like a cell membrane charges are not shielded by non-polar molecules and can have an effect over distances covering entire proteins. To find a good (drug-) ligand for a protein or DNA surface (a receptor), one only has to study the structure and function of natural agonists and antagonists, or the surface topology of the binding site on the macromolecule. Using structural information for drug discovery is referred to as rational drug design and makes use of the concepts of chemical similarity and complementarity.

Chemical similarity is measured by identifying distances between atoms on a receptor and a ligand. Based on the chemical properties of the interacting atoms (or group of atoms = functional groups) small differences in distance have a great influence on the 'reactivity' of a ligand. Since proteins are fluid like entities (alas highly viscous ones), their structures are very sensitive towards disturbances at their surface. For ligands that are similar but not identical, disturbance results in different molecular properties (antagonist or agonist; inhibitor or activator). Thus a local conformational change initiated by the agonist, but not antagonist binding results in a destabilization of the protein structure.

This destabilization is not strong enough to denature the protein, but results in a long range effect across the protein affecting its active site several angstrom away from the ligand binding site. This is known as an allosteric mechanism. As a rule, agonists induce structure destabilization, while antagonists merely bind, but do not affect the protein structure (or trigger a conformational change that locks a protein in its inactive position).

One way to visualize the action of ligands on receptors is to realise that proteins constantly undergo conformational changes which is best described as an equilibrium between an active and inactive, or even among multiple states, including desensitized states (different types of inactive states). Agonists and antagonists shift this equilibrium towards an active or inactive conformation, respectively.

The Pharmacophore

In general, using the surface topology of a group of ligands that all exhibit effector quality (agonist or antagonist) can be overlapped and the contours of all molecules averaged into a union surface. This union surface of a ligand is expected to be complementary to the surface mold of the corresponding binding site on the receptor or enzyme. In complex structures the distribution and combination of physical properties used to search for similarity (complementarity) is large. Modeling structures of ligands in different ways and superimposing different structures with similar affinity exposes the critical fragment or overall similarity of these fragments. The critical fragment of an antagonist/agonist structure is called the pharmacophore. The classical stick and ball model of chemical structures allows to overlap the bond structures and identify these critical segments, often single atoms, in ligands.

In many cases ligand receptor interaction is not necessarily mediated by the entire ligand structure, but by ligand points or critical fragments, the pharmacophore. Thus when analyzing existing data of antagonist and agonist structures, it becomes clear why compounds belonging to very different classes of chemicals so often act on the same target proteins.

Select Ligand Systems

Chlorpromazine is an agonist of the dopamine and contains a superimposable fragment. Neurotoxins saxitoxin

(STX) and tetrodotoxin (TTX) block voltage gated sodium channels. A solvent accessible surface area match shows that the dissimilar structure have identical surface topology.

Sigma ligands (steroidal hormone receptor antagonists) show common points or critical fragments, triangle representation of pharmacophore. Benzodiazepine (GABA antagonist) and beta-carboline fit the same surface mold based on the modeling of the solvent accessible binding site topology, embedded ligand points and hydrophobic core.

Monooxygenase P450 substrate (camphor) and inhibitor (phenyl-imidazol). Fragments can be designed for enzyme inhibitors that mimic the structure, but are not hydrolysable. An example are the HIV protease inhibitors where commonly employed bioisosters (non hydrolysable) are replacing the functional amide group (hydrolysable).

Dihydrofolate Reductase Inhibitors

Ligands binding to cell surface receptors are either amphipathic/hydrophilic/charged and often too large to cross the cell membrane. Intracellular receptors are targets for small hydrophobic ligands which easily diffuse across membranes. A well studied model system includes the inhibitors of bacterial dihydrofolate reductase (DHFR). This enzyme catalyzes the reduction of dihydrofolate (DHF) to tetrahydrofolate (THF), an important step in DNA synthesis.

Comparing sequence and structure of DHFR from different organisms shows many similarities but also explains why some inhibitors are selective against bacteria, while having little effect on the enzyme of the host organism. Inhibitors of human DHFR are also studied for their effectiveness as anticancer drugs. Again, the need of tumors to synthesize DNA is much greater than for surrounding healthy cells. DHFR inhibitors therefore may act as chemotherapeutic as well as antibacterial agents.

Serotonin Receptor Agonists and Antagonists

Serotonin (5-hydroxytryptamine or 5-HT) acts on any one of nine known serotonin receptors which play important roles in neuronal signaling in the central and peripheral nervous system.

The cytotoxic agents used in cancer chemotherapy provoke the release of 5-HT from enterochromaffin cells in the peripheral vagal afferent fibres of the gastrointestinal tract initiating vomit reflexes (emesis). 5-HT3 receptor specific antagonists block this action and thereby greatly reduce the number of emetic episodes that occur during cancer chemotherapy. 5-HT3 receptor antagonists have been shown to produce beneficial effects in animal models of cognitive and psychiatric disorders. Whether 5-HT3 receptor antagonists may have similar profound effects in the treatment of anxiety, depression or psychosis will be determined by the outcome of ongoing clinical trials.

However, it is in the treatment of cancer chemotherapy induced emesis that 5-HT3 receptor antagonists have had their greatest impact. The marked clinical efficacy of 5-HT3 receptor antagonists such as ondansetron, granisetron and tropisetron together with their lack of adverse side effects has greatly improved the treatment of cancer chemotherapy induced emesis. 5-HT3 receptors belong to the family of ligand-gated ion channels.

When activated, ions flowing through these channels depolarize the cell membrane triggering action potentials and thus nerve conduction events. 5-HT3 receptor-mediated ion currents evoked by the full agonists 5-hydroxytryptamine (5-HT), quaternary 5-HT (5-HTQ), meta-chlorophenylbiguanide (mCPBG) and the partial agonists dopamine and tryptamine in whole-cell voltage clamp experiments can be used to characterize binding properties of these ligands such as affinity and specificity. Ligand-gated receptors typically switch into an inactive (desensitized) state within seconds of activation.

Both serotonin and its synthetic analogues desensitize the 5-HT3 receptor completely with a steep concentration dependence and a potency order of: mCPBG > 5-HTQ >> 5-HT >> tryptamine > dopamine.

The time course of recovery from desensitization depends on the agonist used. A quantitative molecular pharmacophore model was derived to predict drug affinities for 5-hydroxytryptamine (5-HT3) receptors. The model was based on the molecular characteristics of a learning set of 40 pharmacological agents that had been analysed previously in radioligand binding studies.

Molecules were analysed for various structural features, i.e., the presence of a benzenoid ring and nitrogen atom, substitutions on the benzenoid ring, the location of the substitutions on the nitrogen, and the molecular characteristics of the most direct pathway from the benzenoid ring to the nitrogen. Weighting factors, based on published 5-HT3 receptor affinity data, were then assigned to each of 10 molecular characteristics. The following nine rules have been established for the 5HT-3A receptor pharmacophore structure:

- Contains aromatic ring structure.
- A tropane ring embedded nitrogen is present and located at a nearest distance from the aromatic ring is not more than seven atoms from the aromatic ring
- When aligning the tropane ring nitrogen in the same plane as the aromatic ring (torsion angle flexibility) the distance between the nitrogen and the aromatic ring centre is 6.0 to 7.8Å.
- Chemical substitutions of no more than 3 atoms are allowed at the nitrogen
- The tropane ring structure itself does not tolerate substitutions larger than methyl groups (-CH3); larger groups significantly reduce affinity
- The linker structure between aromatic ring and ring nitrogen contains steric similarities that reduces flexibility (carbonyls or C=O bonds)
- The first and second atom from the aromatic ring in the linker is never a tetrahedral carbon
- The third atom from the aromatic ring may be a tetrahedral carbon
- Substitutions on the aromatic ring must be able to adopt a co-planar conformation

A naturally occurring drug, atropine, demonstrates how a 'small' but significant violation of the 'rules' in its structure explains its very low affinity for the serotonin receptor 3A subtype. The first atom from the aromatic ring is a tetragonal carbon and thus allows torsion angle flexibility explaining that the structure can deviate from being co-planar with the aromatic ring.

These two atoms are the main difference as compared to the highly specific antagonist ICS 205-930. The molecular weight and chemical formula are nearly identical although the affinity of atropine for the 5HT-3A receptor is x1000 lower than that for ICS 205 930. Atropine is a natural antagonist of cholinergic receptors (acetylcholine receptors) and is used as an antidote to nerve gas (sarine) which inhibits acetylcholine esterase. The toxic effect of prolonged acetylcholine stimulation is thus reversed by blocking the acetylcholine receptor.

The case of atropine and the 5HT-3A pharmacophore structure requirements demonstrate the influence of molecular motion on binding. The flexibility of the atropine molecule between its aromatic and tropane ring structures essentially reduces the chance in atropine of superimposing the tropane ring and aromatic ring, a gross-structure requirement for 5HT-3A antagonist binding.

Further *structure-activity relationship studies* (SARS) show that all known receptor antagonists exhibit at least one degree of freedom. This hinders potential screening of ligand-receptor topology which is best achieved with a rigid molecule as template for rational drug design. This importance of molecular motion in ligands (and receptors) is further demonstrated in thermodynamic analysis of drug binding.

Water as Solvent

Dipole structure of water. Water is composed of one oxygen and two hydrogen atoms forming two O-H single bonds of 0.95 angstrom (Å) in length and a bond angle of 104.5° between them.

Based on the asymmetric distribution of electrons in this triatomic molecule, with the electrons attracted to the oxygen nucleus, the water molecule exhibits a molecular dipole moment of 1.84 Debye. A dipole moment m is defined by two point charges q separated by a distance r; m = qr [Cm].

The value of the dipole moment depends on the difference of the electronegativity of atoms sharing a covalent bond structure. The electronegativity series of biologically important atoms (with increasing affinity for electrons) is: H < C << N < O.

DIELECTRIC CONSTANT

Molecular dipoles experience either an attractive or repulsive force and react to external electric fields. This property is known as polarizability of the medium and expressed as dielectric constant D (or e) of the solvent. The dielectric constant determines the polarity of a solvent and thus the solubility of molecules. Polarizable solvents (solutes) are polar or hydrophilic (liking water; water is a polar solvent), while non-polarizable solvents (solutes) are non-polar or hydrophobic. As a general rule, hydrophilic solvents mix well with hydrophilic solvents (solutes), and hydrophobic solvents with hydrophobic solvents (solutes).

HYDROGEN BOND

The dominant electrostatic interaction in water, based on its permanent molecular dipole moment, is the hydrogen bond (H-bond). The hydrogen bond is stronger than an induced dipole-dipole interaction. The latter is known as Van der Waals interaction, a small electrostatic attraction. The hydrogen bond, however, is weaker than a covalent bond. The relative strength of these three types of bonds can be directly assessed by comparing the length of each bond; O-H covalent bond = 0.96 Å (strong), O-H hydrogen bond = 1.8 Å (medium), and O-H Van der Waals bond = 2.6 Å (weak).

Based on the tetrahedral bond architecture and the orientation of two unpaired electron pairs on the oxygen atom, water molecules can form as many as four (4) hydrogen bonds with each other. This maximal extend of hydrogen bonds, or saturated hydrogen bond network, is achieved in water's solid state - ice crystals. Liquid water has an average of 2.3 hydrogen bonds per molecule. The system is highly dynamic, the lifetime of an hydrogen bond is very short, and as a consequence there is no discernible structure in liquid water.

Hydrogen bonds can also be formed by amine groups containing N-H single bonds or carbonyl bonds (C=O). The ability of water molecules to form hydrogen bonds with themselves and biological (macro-) molecules is the single most important parameter to understand structure, function, and regulation of enzymes, genes, and biological membranes. +

IONS AND MOBILE CHARGES

Water is rarely a pure solvent, but contains a multitude of salts, which all exist in the form of dissociated, charged ions. Table salt NaCl, quickly and spontaneously decays into Na+ and Cl- ions. This process is driven by a change in heat capacity of the system, an enthalpic reaction, and at the molecular level is stabilized by the formation of hydration shells.

Hydration shells are semi-stable structures of water molecules that interact with their dipole moment to the central point charge more strongly than they do with themselves. The positive and negative point charges function as external field with the electric field of the dipole moments reorienting against the charge field to minimize the free energy of the system.

The strength of electrostatic interactions between ion pairs like NaCl is described by Coulomb's law, which says that the force holding two equal, but opposing charged ions together is a function of the charge itself, the inverse square distance between the charges and the dielectric constant of the medium. The kinetic energy prevents the durable formation of hydration shells. The ease of solubilization depends on the polarizability of the solvent molecule, a parameter that is a function of the dipole moment as well as the mass and rotational lateral mobility of the molecules.

HYDROPHOBIC EFFECT

Many biologically relevant molecules are partially hydrophobic, meaning that they are not easily soluble in water, because they lack the ability to form strong electrostatic interactions or hydrogen bonds with the solvent. Hydrophobic interactions are typically based on Van der Waals forces (induced dipoles). Their inability to form energetically favourable interactions with water molecules (hydrogen bonds) induces phase separation.

Water molecules preferentially interact with each other through hydrogen bonds. Since hydrophobic molecules must form contact with water molecules, but can do so only through Van der Waals forces (weak forces), every water-solute interaction is thermodynamically less stable than corresponding hydrogen bonds (strong forces) among water molecules themselves. The reduced number of potential hydrogen bonds found on hydrophobic surfaces reduces the degree of freedom of water molecules at these interfaces.

Rotational and lateral movements are restricted and stable water structures are formed at interface boundaries. Reducing the total area of the hydrophobic surface is energetically favourable. Such a reduction is achieved by clustering hydrophobic solutes into large aggregates. The large number of water molecules no longer needed to form less favourable Van der Waals bonds with the hydrophobic solutes, increases the entropy of the system. The entropy of the liquid water phase (rotational, vibrational, translational degrees of freedom) dominates the thermodynamics of the system.

The increase in entropy of the water phase is much larger than the loss in entropy of the aggregated (structured) hydrophobic particles. This entropy driven aggregation of hydrophobic molecules in aqueous solutions is called the hydrophobic effect. It is the major stabilizing force in biological systems determining such wide ranging processes as protein folding, ligand binding, and cell membrane formation.

Binding Energetics

For the search of new and effective agonists and antagonists, computer modeling has become an invaluable tool because powerful processors readily calculate the properties necessary to define a chemical similarity space. Not only can they be used to design new structures, or modify known structures of agonists/antagonists, they are also useful to screen existing compound libraries for structural and chemical similarity.

As it turns out, molecular modeling tools are better at simulating specificity (conformation) than affinity (energy) of interactions. There is a simple reason for it and it has to do with the solvent. Molecular modeling (static or dynamic) is usually performed 'in vacuum' drastically reducing the number of calculations by excluding solvent-solvent and solvent-solute interaction. Missing from theoretical analysis of molecular interaction is the role that surface bound water molecules play during the formation of a ligand-receptor complex.

The role of these water molecules helps explain experimentally observed affinities for naturally occurring ligand-receptor systems that can not be explained by analyzing the non-covalent interaction between ligand and receptor surface in the complex alone. The challenge here is to understand the binding energy components enthalpy and entropy during complex formation for both ligand-receptor binding and solvent displacement.

$$R + L = RL$$

$$Ka = [RL]/[R][L]$$

Both ligand (L) and receptor (R) come in hydrated form and ligand-receptor (RL) complex formation requires replacement of surface bound water molecules from both the ligand and receptor binding site (partial dehydration). Often, surface bound water is more structured than liquid bulk water. Thus, the release of many water molecules upon complex formation into the bulk phase increases the entropy of the entire system (protein-ligand solution). Such an entropy driven process is well described for hydrophobic and amphipathic solutes and is known as the hydrophobic effect.

The Enthalpy-Entropy Compensation Challenge

While it has been found that there is little correlation between the change in Gibbs free energy (DG) of binding and the change in solvent accessible surface area, experimental observation show that despite the overall small change in Gibbs free energy of binding, both its enthalpic and entropic component can be large, yet in opposing direction. Unfavourable enthalpic components of dehydration and ligand-receptor binding can be offset by favourable entropic components stabilizing the ligand-receptor complex.

$$DG = DH - TDS = -RTlnKa$$

Generally, increased bonding in a bimolecular interaction will produce a more negative enthalpy change, DH, but this will come at the expense of increased order associated with a more negative entropy change, DS<0. The inverse relationship observed between enthalpy and entropy changes in binding interaction is known as the enthalpy-entropy compensation. Overall, favourable entropy terms of partial dehydration of ligand and receptor binding site offset the unfavourable entropic term of the more ordered ligand-receptor complex.

This generally explains how living organism can form and maintain ordered structures - create order out of chaos - at the expense of environmental energy. In terms of rational drug design the enthalpy-entropy compensation is a difficult challenge that must be overcome to significantly improve the prediction of binding affinity of novel drugs to novel targets. Nevertheless, successful design of drugs on enzymes with deep binding pockets occluding bulk water (i.e. ordered water structure in binding pocket, a favourable condition for high affinity binding) has been achieved.

Examples of drugs are Nelfinavir and Ro 46-6240. Non-peptidic, small-molecule mimics as inhibitors of protein-protein interaction have proven more difficult to design. Much of the solvent occlusion of peptide inhibitors is provided by the main-chain and Cb atoms (amino acid side chain carbon) adjacent to binding hot spots, which explains why side-chain modifications heave little effect on affinity.

Molecular Crowding

The equilibrium constant Ka used in thermodynamic analysis is a ratio of product over substrate concentration. However, this is an approximation valid only in ideal solutions that do not assume molecular interactions which is equivalent to extrapolating experimental measurements to zero concentration. Concentrations of proteins in cells, however, can be as high as 300 to 400 mg/ml. The correct solution to thermodynamic equilibrium therefore uses a correction called the activity coefficient g.

Multiplying the concentration c with the activity coefficient to correct for real size molecules with real interactions is given as effective concentration or thermodynamic activity, a = gc. The association constant for protein dimerization in bacterial cytoplasm is 8 to 40 fold increased as compared to an ideal solution, while the association constant, i.e., the affinity. Molecular crowding favours association, protein folding, and ligand-receptor (or substrate enzyme) formation.

Binding, however, is also a function of diffusion in solution. Thus, small molecules have favourable diffusion rates even in crowded solutions, while macromolecules experience a drastic drop in diffusion (consider the exclusion volume as in a gel matrix) reducing the positive effect of crowding on affinity.

BIOLOGICAL MACROMOLECULES

Three major types of macromolecules are found in biological systems: proteins, nucleic acids, and carbohydrates (polysaccharides). All play important roles in the physiology and structure of organisms. Catalytic and regulatory functions are mainly performed by proteins, although some ribozymes, RNA based catalytic units act as enzymes. All three types can function as receptors.

Examples are cell surface receptors (proteins) as hormone or neurotransmitter receptors, transcription factors as regulatory elements of gene expression, or glycolipids and glycoproteins as cell surface matrix that is usually cell type or organism specific (pathogenic microorganisms).

Proteins

The role of proteins in cells is three fold; catalyzing

chemical reactions (enzymes); promoting structural stability and mobility (structural proteins and molecular motors), transport of molecules and signal events across biological membranes and filamentous protein structures (e.g. cytoskeleton). Most drug targets are proteins because of their functional importance.

Protein Structure

Proteins are linear polymers of amino acids. There are 20 different amino acids based on their side chain chemical and physical properties. Besides the side chain, every amino acid contains an amino group (NH_2) and carboxyl group (COOH) and a hydrogen atom linked through the central alpha carbon (Ca).

In a protein, the acid-base property of amino acids is not important except for its N- and C-terminal ends, which are always charged at physiological pH values (pH = 7 to 7.5). In the linear polymer, the amino and carboxyl group are covalently linked to form a peptide bond.

Every amino acid residue (except the terminal units) lies at the centre of two peptide bond structures (amide planes) linked by two single covalent bonds. These covalent bonds have rotational flexibility (degree of freedom) and are called torsion angles. The amino acid sequence of a protein is referred to as primary structure (1°D) and largely determines the three dimensional structure (tertiary structure or 3°D) of a protein. The tertiary structure contains repetitive elements dubbed secondary structures (2°D).

This secondary structures are recurrent elements in proteins and can be classified according to the particular polypeptide backbone fold and measured by their torsion angle values. The two most widely found secondary structures are the right handed alpha helix (a-helix) and the beta strand (b-strand). Most active proteins are found in complex with other proteins. The structure of these multi-subunit protein complexes is referred to as quaternary structure (4°D). Protein complexes give the cell an extraordinary functional variability and control over catalytic processes.

The complexity of living organism is achieved not only by the number of different proteins or molecules in general, but by their use as small multi-subunit complexes. Thus the expression of one gene is in many ways dependent on the expression of other genes.

VISUALIZATION OF PROTEINS

Structure Determination Techniques

To obtain information about the structure of molecules at a resolution where the position of individual atoms can be seen, two techniques are currently used: X-ray diffraction from protein or DNA crystals, or nuclear magnetic resonance (NMR) of proteins or nucleic acid in solution.

Although both techniques are used to obtain the same high resolution structure information, they measure completely different physical properties of molecules. X-ray diffraction yields information about the electron distribution in a crystal lattice, while NMR measures the magnetic spin resonance of selected isotopes.

High resolution structures contain detailed information of atoms separated by 1.5 to 3 Å, roughly the length of a hydrogen bond, or twice the length of a covalent bond. NMR: Because of its wide distribution in biological macromolecules, proton-NMR is used to determine the atomic neighbourhood of protons in macromolecules. Currently, NMR solution structures are limited to molecules with a molecular weight smaller than 20 to 25 kDa (for a protein this is about 200 amino acids). NMR is a technique of choice, if the dynamical aspect of structures have to be determined because the time scale of obtaining the data is very short and lies in the order of milliseconds.

X-ray Crystallography

X-ray diffraction measures the electron distribution in crystals that contain millions of units in an ordered structure (crystal lattice and unit cell). The regularity of the crystal lattice determines the level of resolution (the more ordered, the higher the resolution). The sampling of the diffraction data requires much longer times than NMR measurements. Crystal structures are therefore also known as frozen structures. Highly flexible protein domains (disordered regions) often create a 'blank stretch' in the elucidated structure.

Visualization Protein Structure

There exists no unique way to represent the structure of a protein. The 'structure' we choose to describe a molecule depends on the quality or property of the molecule to be studied. For the same reasons that we have physical, chemical, or biological sciences, we have reasons to focus on different structural aspects of proteins. To mention a few, we can choose the simple space-filled model representing the 'shape' or form (Van der Waals surface) of a protein including every atom in the structure, or we can strip the protein to the reduced information of its polybackbone conformation, the most often used representation of protein structures.

We can mark regions of different secondary structures by symbols (cartoon representation) to quickly allow an overview of motifs and domain organization.

To obtain an understanding of the structure-function relationship of a protein, we need to include the physical and chemical properties of its surface and represent the protein in a way that 'makes sense from the point of view of an other molecular entity', in other words its electrostatic behaviour.

This includes clusters of fixed charges, distribution of permanent and induced dipoles, molecular orbitals of aromatic amino acids, the distribution of polar and non-polar amino acids, and the flexibility of surface structures. Their are four major representation of molecular surfaces: the Van der Waals surface, surface potentials, the solvent accessible surface, and the union surface.

Nucleic Acids

DNA as carrier of genetic information may be a target for drug interaction because of the ability to interfere with transcription (gene expression preceding protein synthesis) and DNA replication, a major process in cell growth and division. DNA replication is central to tumorigenesis and pathogenesis. Nucleic acids are not commonly used as drug targets except antisense drugs (RNA) and antimicrobial or anticancer drugs that readily damage DNA strands or prevent regulatory proteins from binding.

DNA Structure

DNA is a linear polymer made of four different types of nucleotides. Nucleotides are complex structures of a cyclic aromatic base, a ribose sugar unit, and one, two, or three phosphate groups. They are named after their different bases, the variable components of nucleic acids, which come in two basic versions - single ring forms called pyrimidines, and double ring forms called purines.

The pyrimidines include cytosine (C) and thymine (T; uracil (U) in RNA), the purines include adenine (A) and guanine (G). The stable form of DNA is a dimer and its tertiary structure the right handed double helix, called B-DNA, or Watson&Crick double helix, named after their co-discoverers. The stability of the B-DNA is provided by the base stacking of flat aromatic ring structures, as well as the hydrogen bonding in base pairs.

In B-DNA only two base pair combinations are found - AT pairs with two hydrogen bonds and GC pairs with three hydrogen bonds. The number of hydrogen bonds and thus the thermodynamic stability of a DNA double helix is directly related to its GC content, i.e., the percentage of GC pairs in DNA. Chromosomal regions with high GC content correlate with the presence of functional genes. There are three principally different ways of drug-binding.

First, through control of transcription factors and polymerases. Here, the drugs interact with the proteins that bind to DNA.

Second, through RNA binding to DNA double helices to form nucleic acid triple helical structures or RNA hybridization (sequence specific binding) to exposed DNA single strand regions forming DNA-RNA hybrids that may interfere with transcriptional activity.

Third, small aromatic ligand molecules that bind to DNA double helical structures by :

- Intercalating between stacked base pairs thereby distorting the DNA backbone conformation and interfering with DNA-protein interaction
- The minor groove binders. The latter cause little distortion of the DNA backbone. Both work through non covalent interaction.

Carbohydrates

Although carbohydrates or polysaccharides play a major role in cell surface recognition, they are not commonly drug targets because they have no enzymatic function but serve as structural components between cells of multicelllar organisms and pathogenic organisms and their hosts. As is the case for nucleic acids, antimicrobial activity of potential novel drugs may include those interfering with binding of pathogenic microorganisms to host cell surfaces.

Both host and microorganism have carbohydrate coated surfaces. The potential of polysaccharide targets for novel drugs is supported by the observation that both microbial DNA and polysaccharides are immunogenic. Thus novel drugs may be developed that mimic an immune response to develop vaccines or produce competitive inhibitors that can interfere with binding.

Recognition of Proteins

Molecular Motion and Protein Stability

Proteins would not function if their structures were not flexible. The flexibility originates from thermal motion of its atoms, the result of their kinetic energy. In living cells and organisms, macromolecular structures are not rigid entities, but resemble highly viscous fluids. As a result, protein activity is temperature sensitive, with too low or too high temperatures causing inactivation.

While low temperatures inactivate proteins because their structure gets frozen or adopts a crystalline state, elevated temperatures cause proteins to unfold or denature. Both conditions compromise the structural integrity of active sites and binding sites and thus reduce activity.

Molecular Dynamics

Protein flexibility is extremely difficult to study experimentally. The time scale of molecular motions ranges from femto- to microseconds, with 'longer' time scales correlating with larger structures and longer distances involved (e.g. protein folding). Modern day computational power enables simulations of thermodynamic flexibility of atoms of larger and larger molecular structures. This computer based simulations are known as molecular dynamics simulations.

Protein Folding

Proteins are synthesized in cells in a linear fashion (on ribosomes) and have to fold into a native, active conformation (tertiary or quaternary structure). This conformation is largely determined by the amino acid sequence and particularly by the distribution pattern of hydrophilic and hydrophobic amino acid residues.

As a general rule, hydrophobic residues are buried inside the protein core during the folding process, driven by the hydrophobic effect. The folding process is temperature sensitive and promoted by the molecular crowding conditions inside cells.

Induced Fit Binding

It has long been established that enzymatic reactions undergo a series of steps called reaction step intermediates. An early intermediate in every reaction is the stabilization of the enzyme-substrate complex, called transition state. The transition state mechanistically explains the ability of enzymes to lower the activation energy of a reaction, thus greatly increasing its catalytic rate. For non-enzymatic reactions, i.e., ligand binding events on receptors, an analogous enhanced interaction between protein and ligand is found.

Here, the initial binding event induces a small conformational change, which increases the molecular closeness between protein surface and ligand, and thus strengthens the interaction. This mechanism is called the induced fit model of ligand binding. At low temperature, i.e. rigid protein structures, ligands loose their affinity for the receptor. Protein folding can be extended to the study of functional changes in protein structures that brings a protein from an inactive to an active state.

Sometimes these conformational changes can be substantial as is the case for the calcium sensing protein calmodulin (troponin C in muscle). Calmodulin undergoes a reorientation of its two domains upon binding of four calcium ions that results in the exposure of a hydrophobic cavity which allows calmodulin to bind hydrophobic target peptides. These target peptides are surface loops on proteins that are inactivated by these loops (self-inhibition).

Calmodulin binding releases the inhibition thus activating the targeted protein/enzyme, often kinases and the calcium pump, a plasma membrane protein responsible for the excretion of cytoplasmic calcium after a physiological signaling event occurred.

Protein Complex Formation

Structural proteins and molecular motors are typically found as large protein complexes and the conformation of the supra-molecular structure is a function of the strength of interaction between protein surfaces. A typical example is the cell-cell adhesion mediated by protein interaction. The neuronal cell adhesion molecules (NCAM) mediate cell contact in a supramolecular complex called cadherin zippers.

The protein-protein interaction is mediated by immunoglobulin like domains and depends on calcium as stabilizing co-factor. These protein-protein interactions are sensitive to environmental conditions like salt concentration, pH, hydrophobicity, temperature or pressure. Molecular motors are protein complexes undergoing controlled changes in their supra-molecular (quaternary) organization (rotation, lateral movement, contraction) due to local changes in cellular electrochemical conditions. Membrane proteins mediate substrate transport or signaling across cell membranes.

Transport is mediated by ion channels, transporters, and pumps. The latter are distinguished kinetically (transport rate) from ion channels which promote a fast, diffusion controlled flux, while pumps control an energy dependent 'uphill' transport.

Pumps regenerate the chemical potential stored by biological membranes and dissipated by ion channels. In the process of dissipation chemical energy is converted into chemical (ATP synthase) or mechanical energy (molecular motors). Although metabolic and membrane transport processes occur under non-equilibrium conditions, they are studied experimentally at chemical equilibrium.

The first step in a catalysis, complex formation, or transport process is a binding event, which is quantified by its equilibrium constant known as dissociation constant, KD, and measures the affinity of a substrate for an enzyme, a ligand for a receptor, a permeant for a transporter. Membrane proteins cannot be understood without an understanding of the structure and function of biological membranes, also known as phospholipid bilayers. Membranes are an example of complex self-assembly systems. Complexity and self-assembly have become important paradigms in modern biology and thus discussed here in some more detail.

The complexity of cellular structures is obtained by arranging molecular components in regular, repetitive arrays. The determining factors of this assembly is the hydrophobic effect, or more generally, phase separation behaviour. The surface structure and shape of the unit molecule defines the overall architecture of the supramolecular structure; its size, shape, and number of units. Unlike macromolecules, which are true polymers and linked by covalent bond formation among units, supra-molecular structures are stabilized by non-covalent bonds.

Examples are Manifold

A very important and intriguing biological supramolecular structure is the cell membrane - a double layer (bilayer) of phospholipids. Phospholipids, the building blocks of cell membranes, are amphipathic molecules, i.e., they are not entirely hydrophobic. They contain a hydrophilic and/or charged headgroup linked to two fatty acid tails. Membranes are stable in water because the hydrophobic fatty acids are protected inside the membrane bilayer by the hydrophilic surface of the tightly packed headgroups.

Cell membranes are perfectly water soluble, and they provide a hydrophobic barrier for small polar/charged molecules. Cell membranes allow compartmentalization of cellular processes.

The hydrophobic barrier, which is essentially an electrical insulator (capacitor) is regulated by membrane proteins that promote transport processes; ion channels for passive diffusion of small ionic species, facilitators and transporters for specific, passive transport, which may be coupled to symport or antiport of a second molecular species (flux coupling), and finally pumps, active transporters that utilize chemical energy in form of ATP hydrolysis or light (photosynthesis).

Thus biological membranes not only form specialized cellular compartments for various metabolic purposes, but also function as storage devices for electrochemical potentials (ion gradients). Other examples of biological self-assembly complexes include ribosomes and chromosomes, large multi-subunit particles of proteins and nucleic acids, the cytoskeletal fibres - microfilaments made of actin and microtubules made of tubulin.

These elongated fibres have two functions. They determine the shape of mammalian cells and they are dynamic systems providing a way of intracellular transport by means of subunit shuffling between fibre ends.

Cells are not a homogeneous solution of molecules, but highly organized compartments. These properties are particularly apparent during embryogenesis, where cells or cell ensembles gain precise polarity, a functional asymmetric distribution of cellular components necessary for proper cell growth and differentiation. The self-assembly properties of small, amphipathic molecules is utilized to design novel, supra-molecular structures with defined functional properties.

The goal is to produce molecular scale structures, molecular motors, fibres, conducting elements etc. in the nanometer range. The technology of producing these tiny assemblies is commonly referred to as nanotechnology. Nanotechnology is an attempt to control the crystallization process of small molecules of varying size, shape and solubility properties.

The formation of cylindrical structures called nanotubes form microscopic channels and molecular sieves controlling transport processes across biological membranes. These nanotubes may provide useful for the design of drug delivery devices. Their specificity for what is transported and across membranes of which cell types could be used to deliver molecules to tumor cells only and not healthy tissue.

Allosteric Properties of Proteins

Allosteric properties are the result of conformation changes induced by molecular interactions with macromolecules by both small ligands and protein-protein interaction.

The conformational changes induced by binding are the essence of regulating the activity of proteins by shifting these macromolecules between functional and non-functional states. Allostery and cooperativity. Often protein (complexes) contain more than one binding site for more than one type of ligand or substrate.

Proteins have the ability to coordinate what is going on at those different binding sites in such a way that the binding of one ligand alters the affinity for an other ligand on the same protein (-complex).

This is an allosteric mechanism. The 'interacting' ligands are not identical. For identical ligands, the allosteric mechanism is called a cooperative effect. Examples of allosteric mechanisms are the ligand binding induced changes in conformation of cell surface receptors (signal transduction) or ion channels (action potential), or the interaction of ligand binding induced dimerization of transcription factors (nuclear receptors). The latter are nuclear proteins, which undergo a change in DNA binding affinity as a function of dimer formation.

The DNA binding event activates or suppresses gene expression or replication activity. A well known cooperative effect is the binding induced increase in oxygen affinity on the four identical binding sites of hemoglobin. While completely devoid of ligand, the affinity for molecular oxygen is very low.

Hemoglobin, which is a tetrameric protein complex with four identical heme binding sites, undergoes a substantial conformational change after the interaction of one molecule of O_2. The conformational change drastically increases the oxygen affinity for the remaining three binding sites, a positive cooperativity between the four binding sites.

Receptor Signaling

The scope of drug targets is as large and broad as the proteins and nucleic acids found in cellular metabolism. There are, however, preferred targets and they are mostly located at the extra cellular surface of cells. Similar to hormones, neurotransmitters, growth factors and natural toxins, many drugs bind to membrane proteins that belong to two major classes - G-protein coupled receptors (GPCR; metabotropic) and ion channels (ionotropic).

The latter are sensitive to local anesthetics, which directly bind to ion channel proteins, while both iono- and metabotropic receptors are sensitive to general anesthetics, which are believed to function through modification of the physical properties of cell membranes.

A special group of membrane interacting antibiotics are pore forming peptides like alamethicin, gramicidin A, and mellittin. They kill cells by perforating the electrochemical gradients of membranes and depleting their energy storage. An estimated 30% of all currently approved drugs bind to G-protein coupled receptors or GPCRs. Although no high resolution structure for this class of receptors is yet available, structural models based on their homology to bacteriorhodopsin, a bacterial proton pump, and bovine rhodopsin, a light sensitive G-protein coupled protein were used for rational drug design for GPCR ligands.

GPCRs are have a simple generic structure with a large N-terminal domain facing the extracellular side of the cell membrane followed by seven transmembrane spanning (TM) domains (alpha helices) and a C-terminal, cytoplasmic domain of variable length. The C-terminal domain and the loop connecting TM5 and 6 interact with G-proteins which are activated by GTP binding and receptor-ligand interaction in what is known as a ternary complex.

P2Y, a purine receptor, has recently been modeled for its structure-activity relationship of ligand docking. Both the receptor binding site and the ligand pharmacophore have been characterized. P2Y1-ATP complex models describe three binding sites: a meta-binding site I, meta-binding site II, and the principal transmembrane spanning segment (TM) binding site. The meta binding sites are provided by extracellular loop structures. A thermodynamic modeling of binding energy indicates that meta-binding site I is almost as strong as the principle TM binding site.

This ligand-receptor docking model does neither include G-protein binding nor molecular motion. In fact, both agonist and antagonist binding are modeled to the same receptor structure.

Recent studies confirm an increasing complexity in receptor activation by dimerization which can lead to changes in ligand specificity and affinity as compared to monomeric receptors. Ligands belong to a variety of chemical structures including small amino acid derivatives (e.g. dopamine) to larger peptides (e.g. opiates). Over 90% of ligands belong to the peptide class.

The following receptor-assisted G-protein activation cycle has been proposed as a two step model:

First, the active receptor ternary complex consists of a receptor with a GDP bound G-protein denoted HRG(GDP).

GDP interaction with the G-protein is the weakest interaction leading to GDP dissociation.

Second, GTP replaces the diphosphonucleotide on the ligand-receptor-G-protein complex. The HRG(GTP) ternary complex is the transition state of the active cycle (consider the ligand-receptor complex as an enzyme that catalyzes GTP loading on G-proteins) and G(GTP) dissociates from the ligand-receptor complex leaving behind an HR complex that can bind a new G(GDP) unit and stimulating nucleotide exchange, while the newly released G(GTP) functions as an effector module on kinases, lipases, ion channels, or adenylyl cyclase. The GPCR system essentially recharges inactive GTPases by accelerating replacement of non-hydrolysable GDP with hydrolysable GTP on Ga subunits.

Ligand-gated Ion Channels

Unlike the G-protein coupled receptors, ligand gated ion channels open an ion selective pore allowing the flow of ions in or out of the cell, depending on the actual membrane potential and ion gradient. These channels serve as receptors for neurotransmitters like glutamate, GABA, glycine, serotonin, histamine and acetylcholine. The receptor for the latter, the nicotinic acetylcholine receptor (nAChR) has been studied pharmacologically, electrophysiologically, and biochemically since the late 1960s.

The kinetics of channel activation and inactivation are well understood and have served as one of the model systems to study allosteric regulation. In this channel, two acetylcholine units bind to each of the two alpha subunits causing the opening of a gate within the membrane spanning portion of the receptor. The ligand binding sites and channel gate are about 25 Angstrom separated and the recent high resolution structural analysis has helped understand the mechanism of this gating or allosteric mechanism.

The gating mechanism of this receptor complex is a nice demonstration of internal structural changes (conformational changes) in response to ligand binding. Here, acetylcholine binding changes hydrogen bond networks within the alpha subunits relaxing some conformational stress within the binding site.

Upon breaking internal hydrogen bonds, the alpha subunit can undergo a conformational relaxation which can affect subunit contact sites tens of Angstroms apart, i.e., the membrane embedded gate. The fact that the channel is a pentameric protein complex underlines the observation that protein-protein interaction within enzymatic complexes allow fine tuning and precise control over the activity pattern of proteins.

Observation from the nAChR also show that protein complexes spontaneously switch between conformational states - active and inactive - even in the absence of ligand. Thus, the ability of proteins to occasionally adopt active structures even in the absence of the agonist (positive stimulus) corroborates the idea that proteins are essentially fluid entities and that ligand binding and covalent modifications (e.g. phosphorylation and other post-translational modifications) simply stabilize proteins in one of at least two thermodynamically stable conformations. The change between an active and inactive state can thus be described by a chemical equilibrium that switches between a state T (tense) and state R (relaxed).

Both states are equally stable (under the right circumstances) and this similarity in stability can easily be explained by the number of subunit contact sites in each state. The transition from one to the other state requires the breaking of some of these contact sites, but an equal number of similar non-covalent bonds are formed 'trapping' the protein complex in either one of two conformations.

Enzymes and Their Inhibitors

Enzymes catalyze the conversion of substrate(s) into product(s). This process can be measured kinetically, how fast a product is formed, and thermodynamically, in which direction the catalysis proceeds. All metabolic reactions are reversible and are defined by their chemical equilibrium where the net formation of a product is zero. Enzyme catalyzed reactions in vivo are usually not at their chemical equilibrium. They have a preferred direction determined by substrate availability as well as being coupled to large energy releasing reactions (exergonic reactions). The latter makes a catalytic reaction de facto irreversible, a common feature of metabolic pathways.

Although enzymes are often involved in the chemical reaction mechanism (covalent bond formation between substrate and enzyme), they are not chemically modified at the end of the process.

Enzyme catalyzed rates are several orders of magnitude higher than the corresponding spontaneous reaction in aqueous solution. Proteases are enzymes that cleave, or cut, or degrade other proteins by hydrolyzing peptide bonds. This may sound like an uninteresting topic, but protein degradation plays a major role in cellular processes.

Proteases are involved in cellular control mechanisms like the removal of old and unused proteins, an essential part of the turnover pathway of all proteins in the cells and affects cell growth, degradation of proteins and peptides for nutritional purposes, defence mechanisms against intrusive proteins and peptides, or control of protein activity. Being enzymes, proteases can be characterized by their substrate affinity and the catalytic rate of the reaction.

Using proteases to study the effects of single amino acid substitutions (mutations) on catalytic rate and substrate affinity demonstrated that these two properties are linked and that this linkage can be explained by analyzing the conformation of the catalytic or active site of the enzyme.

This analysis showed that four major functional groups are found in the catalytic site of proteases. Chymotrypsin, trypsin, and elastase are three members of the family of chymotrypsin proteases.

They all are serine proteases because the amino acid residue at the catalytic site responsible for the transition state stabilization is a serine. The active site of a serine protease can be divided into four essential structural features required for the catalytic action of serine proteases:

Table: Active Site Structure of a Protease

Structural element of active site	Function
The main chain substrate binding	non-specified binding of polypeptide segment
Specificity pocket	Specific binding of side chains, sequence specificity
Oxyanion hole	Stabilizes transition state S* over S in enzyme
Catalytic triad (Asp-His-Ser)	Forms tetrahedral intermediate; hydrolyzes peptide bond

Proteases have preferential cleavage sites in the sequence of a protein substrate. The specificity pocket provides a small binding pocket consisting of 3 amino acid residues that determine the local polarity and electrostatic potential profile for the interaction of residue n-1 on the substrate on the N-terminal side of the scissile bond.

For the chymotrypsin family of serine proteases we find the following sequence specificities: chymotrypsin binds bulky, aromatic residues, trypsin binds positively charged residues, and the extra cellular matrix protease binds small, non-charged amino acid residues.

The 3-D folds of these three proteases are very similar, although there sequences are not identical, although they are evolutionarily related. Another serine protease family, subtilisin family of serine proteases, are products of bacilli species.

Their overall native fold is very different from that of the chymotrypsin type proteases, but the catalytic dryad conformation is identical. The 3-D fold of subtilisin has an a/b motif (instead of the b-barrel motif of chymotrypsin domains) with five parallel b-strands surrounded by 4 a-helices. The comparison of the two families of serine proteases tells us two different things.

First, it has been reasoned to be an example of convergent evolution, where the formation of a catalytic site has evolved twice, with each serine protease family exhibiting a different overall 3-D structure.

Second, the differences in the 3-D structure gives us an idea of the different cellular locations of the corresponding protein families: the catalytic site of an enzyme is conserved over evolutionary time, while the overall structure is conserved to provide structural stability for optimal activity of the protein in any given environment.

We need only understand that very different sequences can provide similar 3-D structures because water solubility depends only on the distribution of hydrophilic and hydrophobic residues, but not on other chemical properties. Overall structural features thus reflect the location of the protein, if it is located intracellular, extra-cellular, if it is cell membrane protein, or if it is resistant to temperature changes or sensitive to proton or calcium concentrations.

Trypsin (Protease) Inhibitor

Cellular control of proteases is carried out by protease inhibitors. These are small peptides or proteins that can bind to the active site of the protease (competitive inhibitor) but which are not hydrolyzed, thereby blocking the access of substrates, e.g. protecting tissue proteins. One example of a protease inhibitor is the bovine pancreatic trypsin inhibitor (BPTI), a small protein of 58 amino acids.

Its structure has been determined by X-ray crystallography and the protein has been widely used for folding studies of proteins. BPTI binds to trypsin through hydrogen bonding forming a tightly packed interface between inhibitor and enzyme. The Michaelis-Menten constant of BPTI binding Km = 10^-13M. The lysine at position 15 binds to the specificity pocket followed by an alanine. The reaction is blocked at the formation of the transition state intermediate.

HIV Protease

Understanding structure-function relationship of proteins can give vital information for the development of drugs that interact with proteins in a host-pathogen environment. A recent example of rational drug design has been the development of an anti-HIV drug, the protease inhibitor. What is important is that the knowledge of the structure of a protein, which is essential for the life cycle of the virus, has been elucidated by X-ray crystallography and functional studies on related proteases, the aspartate family of proteases, has provided insight into the ligand-enzyme interaction.

Thus, a HIV protease inhibitor has been designed by predicting a structure that binds with an affinity several orders of magnitude higher to the viral protease than to related host proteases. Consequently, virus replication can be inhibited without interfering with of the host metabolism.

The human immunodeficiency virus encodes for an aspartate protease (HIV PR). This protease is essential for proper virion assembly and maturation. Inactivation of this protease has therefore been identified as a therapeutic approach to suppress virus replication and complements already existing drugs interfering with HIV reverse transcriptase.

Essential for the rational design of a protease inhibitor was the successful crystallization of the protease with and without bound substrate. The HIV protease is a member of the family of aspartate proteases and related to the pepsin family of proteases. It is inhibited by pepstatin, the natural inhibitor of pepsin.

The structure of pepsin and its binding of pepstatin are known and this information forms the basis of a successful design of a HIV protease inhibitor by using computer models to identify the best possible inhibitor structure. The active site of aspartate proteases contains a pair of aspartate residues in close proximity with a water molecule hydrogen bonded and oriented optimally to attack the scissile bond of the substrate.

The aspartate pair is located at the domain interface in pepsin, a monomeric protein, and at the subunit interface in HIV protease, a homodimer. The catalytic site in viral and cellular aspartic proteases are very similar, but the importance lies in minute differences in symmetry relations at the interface of domains in pepsin and subunits in HIV protease.

HIV Protease Inhibitor

On the basis of the difference in symmetry at the active site of pepsin and HIV protease inhibitors have been designed that show a much higher affinity for the viral protein than for the host protease.

Ki(HIV) >> Ki(Pepsin)

What happens when an asymmetric substrate (peptide) interacts with a symmetric enzyme? The subunits in the homodimer of the HIV protease are able to distinguish whether they interact with the N- or C-terminal end of the substrate/inhibitor.

Like serine proteases HIV PR contains a specificity pocket for the substrate sequence -P2-P1-P1'-P2'- with residue P1 being either Q or E and residue P2 any hydrophobic amino acid. A good inhibitor exhibits a high affinity for the specificity pocket and contains an non-hydrolysable 'scissile' bond P1-P1'. Substrate analog inhibitors have therefore been designed that function as peptido mimetic.

The scissile amide bond of a peptide substrate is replaced by non-hydrolysable isosteres with tetrahedral geometry (that mimics the substrate intermediate tetrahedral geometry of a peptide substrate).

The binding of a hydroxyethylene peptide mimetic is stabilized by the hydrogen bond formation of the hydroxyl of the backbone with the aspartates in the active site of the protease. The development of protease inhibitors has been accelerated by successfully using the concept that the best inhibitors are those that mimic the transition state structure of the substrate of proteases.

RECOGNITION OF DNA AND RNA

DNA Binding Molecules

Drugs affecting gene expression inhibit the action of hormone regulated nuclear receptors. These are DNA binding proteins which either activate or suppress the transcriptional activity. So called transcription factors are regulated by dimerization induced by ligand binding.

Transcription factor inhibitors either block agonist binding, or prevent dimerization. Examples include the nuclear receptors for estrogen, thyroxin, glucocorticoids, and the morphogen retinoic acid.

In plants the ripening process of harvested fruits can be delayed by inhibiting the expression of a gene involved in ethylene production, the causative agent of ripening. Other drugs affecting gene expression either directly bind to DNA (non-covalent) or chemically modify DNA by cross-linking or strand cleavage. Non- covalent interaction by small non-peptidic molecules is mediated by base intercalation. Aromatic flat molecules integrate themselves between the base pair stacks changing the conformation of the double helix.

Examples include the antibiotics actinomycin D and proflavin. Cross linking agents form covalent bonds mostly to nitrogen groups on guanine bases changing the surface structure of DNA and thus blocking protein binding. Examples include aflatoxin and cis-platin. Other anti-fungal and antibacterial agents induce DNA strand cleavage, such as bleomycin, anthramycin, and tomaymycin, all of which are antibacterial and antitumor drugs.

Nuclear Receptors

Nuclear receptors are transcription factors controlling gene expression activity as activators or repressors. They form a superfamily of currently 69 members and made of seven families.

Based on their ligand specificity, they are split into two groups, type I receptors that bind sterol based ligands (e.g. estrogen, glucocorticoid), and type II receptors that bind non-sterol based ligands (e.g. thyroxine, 9-cis-retinoic acid, vitamin D).

Type I receptors form homodimers, while type II receptors form heterodimers, usually involving one retinoid X acid receptor (RXR) subunit, homodimers, or monomers (steroidogenic factor, nerve growth factor induced gene B). Many novel nuclear receptors being discovered are orphan receptors of type II, meaning that their natural ligand has not been identified yet, although it must be a non-steroid structure based on receptor family classification. Type I and II receptors are activated in different ways. Steroid hormone receptor in the absence of ligand are found in the cytoplasmic compartment complexed with heat shock protein subunits like hsp90, 70, or 56.

Ligand binding causes dissociation of the heat shock subunits, dimerization of the receptor, and transport of the ligand-receptor complex into the nucleus. Type II receptors are localized exclusively in the nuclear compartments and often function as silencer in the absence of ligand by recruiting corepressors. Ligand binding releases the corepressor activating transcription.

Thyroxin

A thyroid hormone affecting metabolic rate, temperature adaptation in warm-blooded vertebrates, regulation of water and ion transport across membranes, regulation of cholesterol metabolism and nitrogen secretion, controls growth rate of mammalian and amphibian cells, is involved in the maturation of the central nervous system, controls amphibian metamorphosis, and regulates some mitochondrial enzymes important in energy metabolism. Thyroxine is synthesized in the thyroid gland, secreted and transported by blood plasma proteins albumin (TBG) or transthyretin (TTR).

Inside cells, thyroxine is bound to cytoplasmic binding proteins (CBPs) such as myocardial myoglobin or thyroxine peroxidases which catabolize the hormone after it is no longer used. The nuclear thyroid hormone receptor TR are encoded by two genes (alpha and beta) and differ in ligand recognition and the effects of ligand in binding coactivators and corepressors. The ligand binding difference is caused by a single amino acid substitution in the binding pocket (Asn in alpha, Ser in beta) of each receptor subtype. Structure and function (thyroxine binding) of TTR are well characterized.

The protein forms a tetrameric complex and binds one thyroxine molecule in a central channel formed of beta sheets. High resolution structures allowed the elucidation of the ligand protein interaction. The most likely interaction are two isosteric conformations. Antagonists to TTR can modulate abnormal growth conditions controlled by the thyroid gland. TTR is also an amyloidogenic protein. Human amyloid disorders. Familiar amyloid polyneuropathy and cardiomyopathy, and senile systemic amyloidosis are caused by insoluble TTR fibrils which deposit in peripheral nerves and heart tissue.

Non steroidal anti-inflammatory drugs have been found to strongly inhibit fibril formation in vitro. The protein-drug interaction stabilizes the native tetrameric TTR conformation. The availability of at least three different natural receptors for thyroxine (T3) allows for a comparative study of agonist and antagonist binding. Usually, different ligand structures are available, but only one receptor structure.

Antisense Drugs

A class of drugs not involving any protein interactions are short synthetic oligonucleotides called antisense DNA or RNA strands. They will bind to either DNA or RNA stretches on chromosomes or RNA blocking gene expression and/or translation. Interestingly, anti-sense drugs have been improved by combining short oligonucleotides with polycyclic intercalating residues on each end drastically increasing affinity of intercalating binding mechanisms while at the same time targeting this intercalating agents to short, gene specific sequences.

Small Molecule Ligand-DNA Interaction

The small ligand drug approach offers a simple solution. The synthesis and screening of synthetic compounds that do not exist in nature, work much like pharmacological ligand for cell surface receptors in excitable tissue, and appear to be more readily delivered to cellular targets than large RNA or protein ligands. The lack of sequence specificity for intercalating molecules, however, does not allow to target specific genes, but rather certain cellular states or physiological and pathological conditions, like rapid cell growth and division that can be selectively suppressed as compared to non growing or slowly growing healthy tissue.

The following properties have been identified as important for the successful modeling of ligand-DNA interaction:

- Degrees of freedom
- Role of base pair sequence
- Counter ion effects
- Role of solvent

This problem is analogous to that of protein ligand interaction. The major requirement for intercalating agents is the planar aromatic ring structure. This structure fits between to adjacent base pair planes and can have some, although much restricted, rotational freedom within the plane of the ring. The ligand itself may have flexibility of structural parts outside the DNA binding site and may contain more than one intercalating sidechain:

The structure of the antibiotic triostin A shows the presence of two quinoxaline (double aromatic rings) units linked through a cyclic peptide structure which is stabilized at its centre by a cystein pair (disulfhydryl covalent bond).

Triostatin A belongs to a family of antibiotics which are characterized by cross-linked octapeptide rings bearing two quinoxaline chromophores. Since the spacing between the chromophores is 3.5A, the intercalation process sandwiches two base pairs between the two quinoxalines.

This phenomenon is called bis-intercalation and has first been described for echinomycin by showing that bis-intercalating drugs cause twice the DNA helix extension and unwinding seen as compared to single intercalating molecule like ethidium. The latter is a chromophor which is activated by UV light and is used by molecule biologists to label nucleic acids in gel electrophoresis or ion gradient centrifugation.

Role of base pair sequence Experimental evidence suggests that base pair sequence does not play a large role on the specific nature of most intercalating complexes. As the structure of triostatin A suggests, however, the linker peptide structure may well promote specific interaction with the DNA surface. The major group specific readout sequence of H-bond donor and acceptor could be involved in triostatin A binding.

Binding site	*GC base pair*	*AT base pair*
W1	H-bond acceptor	H-bond acceptor
W2	blank	blank
W3	H-bond acceptor	H-bond donor
W3'	blank	blank
W2'	H-bond donor	H-bond acceptor
W1'	C-H weak hydrophobic	CH3, strong hydrophobic

While the interaction on the major groove side is distinct for the direction of the base pair (e.g. AT vs. TA), there is no directionality at the minor groove side. Minor groove interaction can, however, distinguish GC content (e.g. TATA box binding protein recognizing AT rich sequences for RNA polymerase initiation complex).

The molecular basis of specific recognition between echinomycin and DNA is due to the hydrogen bonding between the ligand alanine carbonyl groups and the 2-amino group of guanine. This is consistent with the observation that the preferred binding site is the sequence CG.

Counter ion effect DNA is a negatively charged polyanion attracting counter ions, positively charged Na^+, or Ca^{++} and Mg++ ions as well as basic residues of proteins. The presence of small counter ion affect drug binding, since the counter ions can screen and shield the negative backbone surface allowing non electrolytes as well as positively charged ligand to interact more strongly with the DNA target. High ionic strength, however, reduces non covalent interaction mediated by hydrogen bonds and electrostatic interactions.

Role of solvent ligand-receptor binding There are three general classes of interactions that must be considered in solvated ligand-receptor binding :

- Ligand solvent interaction (e.g. hydration shell),
- Receptor solvent interaction,
- Ligand-DNA complex with solvent interaction.

The three classes basically describe the sequence of events of free ligand interacting with its receptor and the change in overall solvent interaction before and after binding. We have seen that the hydrophobic effect is completely described by this system and the contribution of the entropy of free bulk water is the major driving force of hydrophobic ligand receptor interaction. This type of interaction is found in intercalating substrates because the hydrophobic, aromatic side chains interactive favourably with the aromatic environment of the base pair stacking.

The total amount of surface bound water is reduced in the after complex formation. Rational for drug design When a compound intercalates into nucleic acids, there are changes which occur in both the DNA and the compound during complex formation that can be used to study the ligand DNA interaction. The binding is of course an equilibrium process because no covalent bond formation is involved. The binding constant can be determined by measuring the free and DNA bound form of the ligand.

Since many of the intercalating substrates are aromatic chromophores, this can be done spectroscopically. Also, DNA double helix structures are found to be more stable with intercalating agents present and show a reduced heat denaturation. Correlating these biophysical parameters with cytotoxicity is used to support the antitumor activity of these drugs as based on their ability to intercalate in DNA double helical structures.

Improvement of anticancer drugs based on intercalating activity is not only focused on DNA-ligand interaction, but also on tissue distribution and toxic side effects on the heart (cardiac toxicity) due to redox reduction of the aromatic rings and subsequent free radical formation. Free radical species are thought to induce destructive cellular events such as enzyme inactivation, DNA strand cleavage and membrane lipid peroxidation.

Minor Groove Binders

Hairpin minor grove binding molecules have been identified and synthesized that bind to GC rich nucleotide sequences. Hairpin polyamides are linked systems that exploit a set of simple recognition rules for DNA base pairs through specific orientation of imidazol (Im) and pyrrol (Py) rings. The hairpin polyamides originated from the discovery of the three-ring Im-Py-Py molecule that bound to minor groove DNA as an antiparallel side by side dimer.

The optimal goal of polyamide ligand design has been reached with finding structures able to recognize DNA sequences of specific genes. A new strategy of rational drug design exploits the combination of polyamides with bis-intercalating structures. WP631 is a dimeric analog of the clinically proven anthracycline antibiotic daunorobuicin.

This new synthetic compound shows an affinity of 10pM and also showed to be resistant against multidrug resistance mechanisms often encountered in antitumor therapy. Multidrug resistance is a phenomenon where small aromatic compounds are efficiently expelled from the cell by cell membrane transport proteins commonly referred to as ABC transporters.

Drugs that form Covalent Bonds with DNA Targets

- Drugs that interfere with DNA function by chemically modifying specific nucleotides are Mitomycin C, Cisplatin, and Anthramycin.
- Mitomycin C is a well characterized antitumor antibiotic which forms a covalent interaction with DNA after reductive activation. The activated antibiotic forms a cross-linking structure between guanine bases on adjacent strands of DNA thereby inhibiting single strand formation (this is essential for mRNA transcription and DNA replication).
- Anthramycin is an antitumor antibiotic which bind covalently to N-2 of guanine located in the minor groove of DNA. Anthramycin has a preference of purine-G-purine sequences (purines are adenine and guanine) with bonding to the middle G.

Index